Radar Calculations Using Personal Computers: Supplement to Radar Calculations Using the TI-59 Programmable Calculator
William A. Skillman

Artech House, Inc.

International Standard Book Number: 0-89006-141-6
Library of Congress Catalog Card Number: 84-070226

Contents

Preface

There are a number of reasons why I converted my TI-59 programs to BASIC. Although I like the portability of the pocket calculator, Texas Instruments has stopped manufacturing their top-of-the-line programmable calculators so that, eventually, replacement or repair may become a problem. Although Hewlett-Packard is still producing the HP-41 and, in fact, announced an advanced model recently, conversion of my TI-59 programs to *Reverse Polish* would be a sizeable task with little advantage in speed or capability. The appearance of BASIC speaking pocket computers seems to indicate the eventual demise of the old-fashioned programmable pocket calculator.

At the same time the *personal computer revolution* was sweeping the land, so it appeared that some form of BASIC machine would be desirable for the long term. I received numerous calls from purchasers of the TI-59 book looking for BASIC versions of the programs. Then the personal computer revolution swept my place of employment and several HP-86s became available in the office. The ease of creating graphics, the 12 place accuracy, and the advanced BASIC functions convinced me to switch over to the HP-86, and I soon found one in my home as well. Since it is many times faster than the TI-59, more difficult jobs can now be tackled. And the power of numerous mainframes is available with a *modem* (and a suitable charge number) and, of course, other goodies, such as *Electronic Mail, Official Airline Guide, et cetera.*

In the process of converting the TI-59 programs to BASIC, a number of goofs turned up, either in the programs or in the text. In addition, several mistakes have been pointed out by book users. Appendix C is a compilation of corrections, clarification of points, *et cetera.*

Because of the relative ease of programming in BASIC, and the much larger memory of the HP-86, it was easy to include numerous enhancements in the BASIC programs. In several programs exact solutions have replaced approximate solutions. Output plots on either printer or plotter save many hours of labor and are much neater than I can plot by hand. These features plus annotation, listing of variables, error trapping, and prompting have increased the size of the programs from about 15000 instructions in the TI-59 to over 250,000 bytes in the HP-86! With a bit of luck and a bit of pruning, these programs conveniently fit on a single disc. In addition to the translated programs, the disc includes a "binary" program, **GDUMP,** required to dump graphics display to the printer, and a master program, **Autost,** which is automatically loaded and executed at machine turn-on and permits the user to select a program from the table of contents of the TI-59 book.

Extensive use of menus, prompting, and input error trapping provide the *anwenderfreunlichkeit* for these programs that I attempted to provide for the TI-59 programs. In addition, the programs are annotated and modularized into subroutines and segments, each with a descriptive header for ease of understanding of the program flow. To ease the burden of translation of these programs to other BASIC dialects, I elected to minimize the use of many of the advanced BASIC features on the HP-86/87. Although extensive use is made of the graphics capability, it is simple to delete the plot subroutine in the translation process. Substitutions must be made for a few of the advanced features (such as arccosine, arcsine), and for the different symbols and commands used in different machines. Because of the growing popularity of the IBM Personal Computer, Appendix B discusses the translation of these programs to IBM PC BASIC, which is a version of the popular Microsoft BASIC. An example of this translation is included to show the similarity of the two BASIC programs. Translation to other BASIC dialects should follow a similar pattern. Translation to the older HP-85 is relatively simple since HP-86 BASIC is a superset of HP-85 BASIC.

No matter how hard we try, there are always a few "bugs" lurking in the programs. So, as with the TI-59, I encourage you to let me know of any bugs that you find.

William A. Skillman
Baltimore, MD
February 1984

Introduction

The programs in this supplement will perform all calculations possible with the TI-59 programs of the original book, plus numerous other calculations that were natural extensions of the programs. Also, many of the programs provide plots either on the CRT (where they may be dumped to a printer) or on a plotter. The plot subroutine is, of course, unique to the graphics capability of the HP-86/87 and will require modification for other computers. In a number of cases, the much larger memory of the HP-86/87, compared to the TI-59, permitted the consolidation of several TI programs into one BASIC program. Thus, there is not a one-to-one correspondence between the two sets of programs. Table 1 shows the TI-59 programs *versus* the BASIC programs.

Although in many cases these programs use the same algorithms as the TI-59 programs, results may differ slightly due to the different number of significant figures of the two machines. In some programs slightly different algorithms are used and in others the test limits for series approximations have been changed to increase accuracy.

A new program, **Autost,** has been added to permit easy access to the BASIC programs without the necessity of memorizing the names of the programs or consulting a list. This program is automatically loaded and run by the computer when it is first turned on. It can be loaded and run manually at any time after initial turn-on. Two versions are given. The first uses the built-in display and input commands. When run, the first version initially presents the user with a list or "Disc Menu" of the chapter titles from the TI-59 book. Entering the number of the chapter and pressing [END LINE] causes one of two actions. If either Chapter 3 or 4 is selected, then program **DETPROB** or **SIG/CLUT** is loaded and run immediately (using the CHAIN command). If any other chapter is selected, then a new menu is displayed, this time with the titles of the TI-59 programs in the selected chapter. Again, a number is selected, and pressing [END LINE] causes the appropriate program to be loaded and run. At any time, when a program is stopped, another program can be loaded manually by typing LOAD "title" [END LINE], then executed by pressing [RUN]. This is possible since each program is written as an independent or stand-alone program. The second version of the **Autost** program illustrates display enhancements possible with the binary program **UTIL/1,** which is found on the demonstration disc for the HP-86/87. AWRITE is used to position text on the screen instead of DISP. The softkeys are used to position a highlight (inverse video) on the desired chapter or program title and to select the desired one. The TAKE KEYBOARD command locks out all unused keys, so that the selection process is "goof-proof." The RESET key is the only key that has its normal function, so that a recovery can be made in the case of an inadvertent total lockout. Note that whichever version of **Autost** is selected, it must be stored as "Autost" to be recognized by the HP-86/87.

At the start of the execution of each program a menu is displayed which permits selection of a printer to print the input and output data. The first printer selection is the CRT which permits all data, both input and output, to be displayed on the CRT whether a printer is physically connected or not. If this option is selected, no data is sent to a printer. If one of the remaining three printer options is selected, then all input and output data will be printed in addition to being displayed on the CRT. Following printer selection, the main portion of the program is entered and the user is prompted as required for each program. In some cases a menu is presented immediately; in others, initial data is requested. At the end of one complete operation, most programs display a "Main Menu." A number of options are presented, such as entering new data, rerunning the program, or selecting another operation. The last item on the Main Menu is "Disc Menu," which, when selected, causes the **Autost** program to be reloaded and run for selection of another program.

As data is entered, it is checked against limits or "error-trapped" wherever possible. Thus, if a menu is displayed with items numbered 1-6, the prompt "Enter #?" is displayed, and the user enters "7," the computer will BEEP, a message "Use # 1-6" is presented, and the prompt "Enter #?" will be displayed again. Inputs that determine array sizes are checked against the Dimensioned size, and appropriate error messages are displayed if the allocated dimension is exceeded. If the user enters alpha data (or no data) when numeric data is requested, the computer will BEEP, display an error message "Warning 7: NULL DATA." If zero is not an allowed input, then the prompt will be repeated, otherwise execution will continue with the value 0 for the input variable. The computer treats the alpha data as a variable name, but if no value has been assigned to the name, then the variable is NULL. Inputs that will create overflow or underflow errors are trapped in many cases, but no doubt there are some cases which will cause that unnerving "BEEP"! In several programs input data may be reviewed and corrected during or after data entry. In other programs the entire operation must be performed with the incorrect data and then the correct data entered on the second pass. In some cases it may be simpler to stop the program by pressing [PAUSE], and then restarting by pressing [RUN]; or by inspection of the program listing it may be possible to recover by restarting the program at line xxxx by typing "CONT xxxx" and pressing [END-LINE]. (CONT xxxx preserves the current variables, whereas [RUN] or RUN xxxx [END LINE] destroys them.) This same procedure may permit a program to be restarted following a fatal error that stops execution. Most fatal errors are prevented by the error trapping in the programs, but there may be a few cases that will slip out of the tightest trap. Several errors, such as overflow, or underflow will not halt the program, but will BEEP and display an error message, thus disrupting the tidy screen format.

It is very frustrating to see a prompt such as "enter antenna length?" and not have the faintest idea what units the program uses. All prompts in these programs inform the user of the units, except where any units are acceptable (such as where the wavelength and slot spacing of an array antenna are only used to obtain the ratio). In the program **SIG/CLUT**, several combinations of units may be selected for range and altitude, with unit conversions being performed automatically by the program.

Many of the prompts pose a question that requires a "yes" or "no" answer, such as "Dump to printer ?": the proper reply is simply to type a single letter (Y,y,N, or n), then press [END LINE]. Any other reply will get a BEEP and a repeat of the prompt. In the prompts, the abbreviation [EL] is frequently used for [END LINE], which is also called RETURN or ENTER on other computers.

Note that the "default" values mentioned in a few prompts are obtained by pressing [END LINE] without entering any data. The input of this "Null string" provides the proper branching to substitute the default value in place of input data.

The programs in this supplement are all laid out in a similar format. The first line gives the program name and the revision date. Next is a listing of improvements, if any, to the original TI-59 program. These are self-explanatory by either use of the book or by reference to the program itself. The most important variables are then listed in alphabetic order. Dimension statements for arrays are followed by assignment of permanent variables. Defined functions and subroutines, if any, are next, followed by a call to the printer selection subroutine. The printer selection subroutine is usually at or near the end of the program. A subroutine to load the binary program **GDUMP**, which dumps the CRT to the printer, may also be in this location. In general, the main part of the program is divided into relatively small segments with a descriptive header above each segment. The "inner" statements of multiple line FOR...NEXT loops are indented to help show the loop construction. This is also true of loops that use an IF...THEN test to control the looping. Comments appended to statement lines are in lower case to help distinguish from the program itself. To conserve space, both in the program listings and in storage space, liberal use is made of multiple statement lines. In the IF(expression)...THEN...ELSE construction, multiple statements between THEN and ELSE are executed if the expression is true, and multiple statements following ELSE are executed if the expression is false. In many cases the expression is simply a numeric variable which is "false" if the variable is zero, and is "true" if it is non-zero. Note that the ELSE part of the branch statement is optional. Instead of multiple statements, a single statement, or simply a line number (implied GOTO) may be used in the IF...THEN...ELSE branch construction. Although statement labels may be used as branch destinations in the HP-86/87, they have not been used in these programs for compatibility with other BASIC dialects. Although labels are handy in the original construction of a program, they give no clue as to their location, so I prefer to change to line numbers after initial writing and debugging of a program.

For your ease in typing in programs in the auto-numbering mode, the programs have been Renumbered with default spacing of 10. However, some last minute changes have been inserted without renumbering.

The minimum configuration required is an HP-86 with 64K internal RAM, plus a monitor, or an HP-87 with 64K. Several programs are dimensioned to utilize almost all the available RAM, so that the presence of extra BINARY programs may prevent execution with a memory overflow message. All programs except

TABLE I-1
PROGRAM REFERENCE CHART

TI-59 PROGRAM	BASIC PROGRAM
Chapter I Antennas	
Program 1-1 Antenna Parameters	ANTPAR
Program 1-2 Scanned Array Pattern	PATPLOT
Program 1-3 Array Antenna Beam Calculations	ANTBEAM
Program 1-4 Antenna Pattern Synthesis	ANTSYN
Chapter 2 Propagation	
Program 2-1 Radar to Target Angles and Ranges	ANGLES
Program 2-2 Radar Range-Height-Angle Chart, Exponential Atmosphere	TRACE
Program 2-3 Radar Coverage Diagram	COVER
Program 2-4 Low Altitude Target Tracking Error, Monopulse Radar	MONOERR
Chapter 3 Detection Probability	
Program 3-1 Single Look Detection Probability, Marcum & Swelling Targets	DETPROB
Program 3-2 Single Look Detection Probability, Marcum N=1	DETPROB
Program 3-3 Single Look Detection Probability, Chi-Square Target	DETPROB
Program 3-4 Simplified Detection Calculations	DETPROB
Program 3-5 Multilook Detection Probability for Fluctuating Targets	DETPROB
Chapter 4 Signal-To-Noise Calculations	
Program 4-1 Load Radar Parameters and Surface Clutter Model	SIG/CLUT
Program 4-2 Land and Sea Clutter-to-Noise	SIG/CLUT
Program 4-3 Rain and Chaff Clutter-to-Noise	SIG/CLUT
Program 4-4 Range Performance in Clear and Jammed Environment	SIG/CLUT
Chapter 5 Filters and Filtering	
Program 5-1 Digital Filter Analysis	DIGFILT
Program 5-2 Butterworth Digital Filter	DIGFILT
Program 5-3 Chebyshev Digital Filter	DIGFILT
Program 5-4 Fast Fourier Transform	DFT
Program 5-5 Discrete Fourier Transorm	DFT
Program 5-6 Dolph-Chebyshev Weight Coefficients and DFT Response	DOLPH
Program 5-7 Paired-Echo Analysis	PAIRECHO
Program 5-8 Taylor Weighting Coefficients	PAIRECHO
Program 5-9 FIR Canceller Response, Stagger PRF	FIRCANC
Program 5-10 Maximum Entropy Method (MEM) Spectrum	MEM
Program 5-11 Improvement Factor for DFT Filter Bank	IMPACT
Program 5-12 Improvement Factor for FIR Clutter Canceller	FIRCANC
Chapter 6 Receivers and Processing	
Program 6-1 Noise Figure and Noise Temperature	NOISEFIG
Program 6-2 Autocorrelation Function for Sampled Data	AUTOCOR
Program 6-3 Ambiguity Function for Phase-Coded Signals	AMBFUN
Program 6-4 Multiple PRF Ambiguity Resolution	CRT

for **DETPROB SIG/CLUT** and **DFT** should run in a 32K HP-87. More RAM will permit increasing the dimensions of arrays, especially in program **DFT.**

Peripheral options include a line printer and graphics plotter. Provisions are made in the programs to use several versions of Epson printers. The Hewlett-Packard printer HP-82905B, which is a modified Epson printer, can be used as well as the Epson MX-80, MX-100 or FX-80 (designated MX-100 in the printer selection menu of the programs). The Epsons are the same as the HP-82905A, as far as the **GDUMP** binary program is concerned. The HP printer can be connected (depending on the option) to either the

Centronics parallel interface (select code 701), the HP-IB bus, HP82937A (select code 301) or the RS-232 serial interface, HP 82939A (select code 10). The programs must be modified slightly if the older HP-82905A is to be used. The printer selection menu could be extended to a fifth sele____ n which would set PN = 4 (to select the correct commands for **GDUMP**) and PRINTER IS 301 (for th_ _IP-IB bus). Use PRINT CHR$(27)&"\$11L" to skip perforation. **GDUMP** will work with them. Check with HP for your particular printer. Note that the later "B" version of the HP-86, and the HP-87, have a built-in HP-IB interface instead of the Centronics interface, so that you need an HP-IB interface instead of the Centronics interface, so that you need an HP-IB interface compatible printer to connect directly to these computers. An RS-232 interface or the HP 82940A GPIO interface can be added to these computers to interface with RS-232 or Centronics compatible printers, respectively.

The plotter assumed in these programs is an HP plotter on an HP-IB interface. Other plotters may be driven from either the built-in Centronics interface or with an added RS-232-C interface module. Appendix D illustrates the use of a low cost printer which can be driven directly from the Centronics printer port of the HP-86A and, of course, could be used with most personal computers since it also can be driven from a RS-232-C port.

The I/O ROM is an option that is used in four programs to slash the zero on the HP printer. If the ROM is not installed, the program bypasses the zero-slashing segment. However, to avoid editing problems, the statements using the I/O ROM commands should be deleted (or converted to comments by inserting "!" following the line number). The Epsons provide a DIP switch to permanently slash the zero. The statements containing I/O ROM commands are found in the following lines:

Program	line numbers
DETPROB	6410,6420,6570,6590
SIG/CLUT	3030,3190,7430,7440
DIGFILT	3090,3100,3230,3330
DFT	3470,3800,4730-4760

Another option is the use of the binary program **FFTq** in program **DFT** to speed up the **FFT** calculation by a factor of about 5. The binary program **FFTg** is found in the Waveform Analysis Applications Pac available from HP. Again, if the **FFTg** program is not available, the statements (lines 2280 and 4932-4950) using the ROM commands should be deleted (or converted to comments) to avoid editing problems.

When typing in or editing programs that use **GDUMP** (or any other binary program), this program must be resident in RAM. Good practice is to load **GDUMP** (using LOADBIN "GDUMP") from the disc at the beginning of an editing session. In normal usage, **GDUMP** is loaded by the program.

A single disk drive, such as the HP 9130A, is required to store and load these programs. The **Autost** program and those programs using **GDUMP** and **FFTg** require the disc with these programs in the default drive, which is normally the disc drive connected to the port labelled DRIVE ZERO, but DRIVE 1 can be selected by executing the statement "MASS STORAGE IS":D701". If the program that is to be loaded is not on the disk, or there is no disk in the drive, or the drive door is open then execution stops with the appropriate error message displayed. Execution can be continued by inserting the required disk, closing the door, and pressing [CONT].

Chapter 1 Antennas
Program ANTPAR

```
10 ! Pgm ANTPAR , Rev 2/05/84
20 ! Derived from Pgm 1-1 of Radar Calc.  By W. Skillman
30 !
40 ! Important variables:
50 !    BWS.....standard beamwidth, Dolph-Cheby array
60 !    BWA.....az beamwidth (deg)
70 !    BWE.....elev    "      "
80 !    GU......unweighted gain
90 !    GD......directive (weighted) gain
100 !   GSLA....peak to sidelobe ratio in azimuth (dB)
110 !   GSLE....ditto for elev.
120 !   H.......antenna height
130 !   LAM.....operating wavelength
140 !   W.......antenna width
150 !
160 ! ************** Defined Functions *********************
170 DEF FNICOSH(X) = LOG (ABS (X)+SQR (X*X-1)) !  Inverse Hyp Cos
180 DEF FNBWS(r) = 2/PI *SQR (FNICOSH(r)^2-FNICOSH(r/SQR (2))^2)
190 DEF FNRD(X) = INT (10000*X+.5)/10000 ! round to 4 decimals
200 DEF FNPDB(X) = 10*LGT (X) ! Power ratio to dB
210 DEF FNIVDB(X) = 10^(X/20) ! dB to voltage ratio
220 !
230 DIM T$[30],PR$[80],AZ$[80]
240 !
250 T$="** ANTENNA PARAMETERS **" @ DR=180/PI
260 AZ$="Azimuth & Elevation Sidelobe levels WRT peak (dB)="
270 !
280 GOSUB 660 ! init, printer sel.
290 GOSUB 420 ! select antenna parameters
300 GOSUB 480 ! select sidelobe levels
310 GOSUB 540 ! calc beamwidths and gain
320 !
330 DISP @ DISP "Main Menu: select next operation"
340 DISP "1   New Antenna Parameters"
350 DISP "2   New Sidelobe Levels"
360 DISP "3   Calc beamwidths and gain"
365 DISP "4   Disc Menu"
370 DISP "Enter # ";@ INPUT R @ IF R=4 THEN CHAIN "Autost"
380 IF R<0 OR R>4 THEN DISP "Use # 1-4" @ BEEP @ GOTO 370
390 ON R GOSUB 420,480,540 @ GOTO 330
400 !
410 ! *************   Enter antenna parameters  **************
420 DISP "Enter antenna parameters using same units of length"
430 DISP "Enter operating wavelength";@ INPUT LAM
435 IF LAM<=0 THEN DISP "Use wavelength > 0" @ BEEP @ GOTO 430
440 DISP "Enter antenna width and height";@ INPUT W,H@ WL=W/LAM @ WH=H/LAM
445 IF W<=0 or H<=0 THEN DISP "Use dims > 0" @ BEEP @ GOTO 440
450 IF PN#1 THEN PRINT "Wavelength = ";LAM;", Width =;W;, Height= ";H
460 WL=W/LAM @ WH=H/LAM @ GU=4*PI *WL*WH @ RETURN
470 !
```

```
480 ! **************   sub to select antenna sidelobe levels   ***********
490 DISP "Enter ";AZ$;
500 INPUT GSLA,GSLE@ GSLA=ABS (GSLA) @ GSLE=ABS (GSLE)
510 BWSA=FNBWS(FNIVDB(GSLA)) @ BWSE=FNBWS(FNIVDB(GSLE))
520 IF PN#1 THEN PRINT "AZ$;GSLA;", ";GSLE
525 RETURN
530 !
540 !***************  calc beamwidths and gain *********************
550 BWA=DR*BWSA/WL @ BWE=DR*BWSE/WH
560 PR$="Azimuth and Elevation beamwidths (deg)= "&VAL$ (FNRD(BWA))&", "
570 PR$=PR$&VAL$ (FNRD(BWE)) @ GOSUB 630
580 GD=GU*.88589^2/BWSA/BWSE
590 PR$="Unweighted and directive gains (dB)="&VAL$ (FNRD(FNPDB(GU)))&", "
600 PR$=PR$&VAL$ (FNRD(FNPDB(GD))) @ GOSUB 630 @ PR$=""
610 !
620 ! ****************   DISP/PRINT   *****************************
630 DISP PR$ @ IF PN#1 THEN PRINT PR$
640 RETURN
650 !
660 ! ***********  INITIALIZATION &  PRINTER SELECTION   **************
670 PAGESIZE 24 @ CLEAR @ DEG @ PRINTER IS 1
680 DISP TAB (10);T$ @ DISP
700 DISP "Select Printer:" @ DISP
710 DISP "1  CRT"
720 DISP "2  HP-82905B--HP-IB"
730 DISP "3  HP-82905B--Centronics"
740 DISP "4  MX-100"
750 DISP "Enter #";@ INPUT PN
760 IF PN<1 OR PN>4 THEN DISP "Use # 1-4 " @ BEEP @ GOTO 750
765 IF PN#1 THEN DISP "Make sure Top-Of-Form is set properly, Enter
date(no',')";@ INPUT DA$
770 ON PN GOTO 850,790,800,820
780 !
790 PRINTER IS 301 @ GOTO 810
800 PRINTER IS 701
810 PRINT CHR$ (27)&"&l1L" @ GOTO 840
820 PRINTER IS 701 @ PRINT CHR$ (27);"N";CHR$ (8)
840 PRINT TAB (15);T$ @ PRINT TAB (25);DA$
850 RETURN
```

Program PATPLOT

```
10 ! Pgm PATPLOT, rev 2/05/84, by W. Skillman
20 ! Derived from Program 1-2, Radar Calc..
30 !
40 ! Improvements vs. Radar Calc...
50 !    Doubled Angle Resolution
60 !    0 to -60 dB Gain Scale (vs 0 to -40)
70 !    CRT or Line Printer plot
80 !
90 ! Important variables:
100 !    CTH..COS(TH) array
110 !    D....element spacing
120 !    F....Gain (dB) array
130 !    LAM..wavelength
140 !    M....# of elements
150 !    STH..SIN(TH) array
160 !    TH...angle from reference(deg)
170 !    THS..scan angle from  "
180 !
190 DIM Y$[60],YN$[80],YB$[80],PR$[80],X$[73],F(73),YD$[70],T$[60]
200 DIM STH(73),CTH(73)
210 !
220 T$="ANTENNA PATTERN-Unweighted array" @ E$=CHR$ (27) @ YB$=CHR$ (124)
230 B$="          " @ Y$=B$&B$&B$&" Relative Gain (dB)" @ C$=B$&YB$
240 YN$=B$&" -60       -50       -40       -30       -20       -10         0"
260 X$=B$&B$&"     A N G L E   D E G R E E S        "&B$&B$
270 GOSUB 790 ! init, printer select
275 YD$="+" @ D$="---------+" @ W$="Please wait while "
280 DISP W$;"I store sines and cosines"
310 FOR I=1 TO 6 @ YD$=YD$&D$ @ YB$=YB$&C$ @ NEXT I
320 I=1 @ FOR TH=-90 TO 0 STEP 2.5 @ CTH(74-I),CTH(I)=COS (TH)
330 STH(I)=SIN (TH) @ STH(74-I)=-STH(I) @ I=I+1 @ NEXT TH
340 !
350 ! **********   MAIN PROGRAM   *************************
360 GOSUB 430 ! enter antenna parameters
370 GOSUB 610 ! plot pattern on CRT or Line Printer
380 ! Reset printer line spacing to normal 12 dots or 6 lines/inch
390 IF PN=4 THEN PRINT E$;"2" ! MX-100
400 IF PN=2 OR PN=3 THEN PRINT E$&"&l6D" ! HP-82905B
410 DISP "More plots";@ INPUT R$@ IF R$="Y" THEN 360 ELSE CHAIN "Autost"
420 !
430 ! **********   Enter antenna parameters   ***************
440 BEEP 20,50 @ DISP "Enter element spacing and wavelength (same units)";
445 INPUT D,LAM @ IF D<=0 OR LAM<=0 THEN DISP "Use #s>0" @ BEEP @ GOTO 440
450 IF D<LAM/2 THEN DISP "Use spacing >= wavelength/2" @ BEEP @ GOTO 440
460 DISP "Enter # of antenna elements";@ INPUT M
470 IF M<= 0 THEN DISP "Use # >0" @ BEEP @ GOTO 460
480 DISP "Enter desired scan angle (deg)";@ INPUT THS@ SAS=SIN (THS)
490 IF ABS (THS)>90 THEN DISP "Use angle less than +/-90" @ BEEP @ GOTO 480
```

```
500 DISP W$;"I calculate gains for the plot" @ IF PN=1 THEN 540
510 IF FLAG THEN PRINT CHR$ (12) ELSE FLAG=1 ! Form feed after 1st graph
520 PRINT B$;T$;"        ";DA$ @ PRINT "        Spacing=";D;", Wavelength=";
530 PRINT LAM;", # of elements=";M;", Scan angle=";THS @ PRINT
540 FOR I=1 TO 73 @ GOSUB 560 @ NEXT I @ RETURN
550 !
560 ! ************ CALC GAIN FOR UNIFORM ILLUM  **************
570 u=180*D/LAM*(STH(I)-SAS) @ R=1
580 Su=SIN (u) @ IF Su THEN R=SIN (M*u)/M/Su
590 E=R*(1+CTH(I))/2 @ E=MAX (EPS ,ABS (E)) @ F(I)=20*LGT (E) @ RETURN
600 !
610 ! ********* Subroutine to plot gain vs. angle  ***********
620 PR$=Y$ @ GOSUB 760 @ PR$=YN$ @ GOSUB 760
630 SN$=" " @ J=1 @ AN=-90
640 ! Set Printer spacing to 8 dots/line or 9 lines/inch
650 IF PN=4 THEN PRINT E$;"A";CHR$ (8) ! MX-100
660 IF PN=2 OR PN=3 THEN PRINT E$&"&l9D" ! HP-82905B
670 FOR I=1 TO 18
680   PR$="        "&X$[J,J]&SN$&VAL$ (AN)&" "&YD$ @ GOSUB 730 @ AN=AN+10
690   FOR L=1 TO 3 @ PR$="        "&X$[J,J]&"        "&YB$ @ GOSUB 730 @ NEXT L
700   IF AN=0 THEN SN$="    "
710   IF AN>0 THEN SN$="   "
720 NEXT I @ PR$="        "&X$[J,J]&"   90 "&YD$
730 K=72+F(J) @ J=J+1 @ IF K>= 12 THEN PR$[K,K]="*"
740 !
750 ! ************* DISP/PRINT  *****************
760 DISP PR$ @ IF PN#1 THEN PRINT PR$
770 RETURN
780 !
790 ! *********** INITIALIZATION &  PRINTER SELECTION  ****
800 PAGESIZE 24 @ CLEAR @ DEG @ PRINTER IS 1 @ FLAG=0
810 DISP B$;T$ @ DISP
820 DISP "Select Printer:" @ DISP
830 DISP "1  CRT"
840 DISP "2  HP-82905B--HP-IB"
850 DISP "3  HP-82905B--Centronics"
860 DISP "4  MX-100"
870 DISP "Enter #";@ INPUT PN
880 IF PN<1 OR PN>4 THEN DISP "Use # 1-4 " @ BEEP @ GOTO 870
890 ON PN GOTO 940,910,920,920
900 !
910 PRINTER IS 301 @ GOTO 930
920 PRINTER IS 701
930 DISP "Set printer to top-of-form, enter today's date (any ";
935 DISP "comma-less format)";@ INPUT DA$
940 RETURN
```

Program ANTBEAM

```
10  ! Pgm ANTBEAM, Rev 2/13/84
20  ! Translated from Pgm 1-3 of Radar Calc. by  W. Skillman
30  !
40  ! Improvements vs Radar Calc:
50  !    Supergain check (element spacing < LAM/2 not allowed)
60  !    Scan angles > 90 deg not accepted
70  !
80  ! Important Variables (all angles in degrees, lengths in cm)
90  !    a.....wide dimension of W/G
100 !    BWMZ..bandwidth (MHZ)
110 !    BWN...Normal beamwidth with spreading factor
120 !    d.....interelement spacing
130 !    DPH...interelement phase shift w/o F
140 !    F.....   "        phase shift w/o phase shifters (0 or 180)
150 !    FLAG..0=Coax, 1=Waveguide  (delay line)
160 !    K.....dielectric constant of delay line medium
170 !    L.....antenna length
180 !    LAM...operating wavelength
190 !    LAMC..center      ''
200 !    LAMCF.cut-off      ''
210 !    n90...value of n for beam at 90 deg (may be fraction)
220 !    S.....delay line differential length
230 !    SENS..frequency scan sensitivity (deg/MHZ)
240 !    SF....Spreading factor
250 !    TH....angle from array normal
260 !    THCM..highest beam angle with 0 phase shift, center freq.
270 !    THi...Beam angle(s)
280 !    THM...max scan angle in coverage
290 !    THS...beam scan angle
295 !    TPH...total interelement phase shift/360
300  !
310 DIM PR$[80],T$[35],DR$[30]
320 T$="**ANTENNA BEAM CALCULATIONS**" @ L$="Wavelength "
325 CF$="center frequency"
330 !
340 ! *************  Defined Function  *********************
350 DEF FNRD(X) = INT (10000*X+.5)/10000 ! Round to 4 decimals
360 !
370 GOSUB 1720 ! init, printer selection
380 GOSUB 620 ! Enter antenna parameters
390 !
400 ! ****************   Main Menu  **************************
410 PR$="" @ GOSUB 880 @ DISP "Main Menu: select next operation"
420 DISP "1  New Parameters "
430 DISP "2  Calc Guide ";L$
440 DISP "3  Scan Calculations"
450 DISP "4  Disc Menu"
460 DISP "Enter #";@ INPUT R@ IF R=4 THEN CHAIN "Autost"
470 IF R<1 OR R>4 THEN DISP "Use # 1-4" @ BEEP @ GOTO 460
480 ON R GOTO 380,500,910
490 !
```

```
500 ! *********  Keyboard calc of Guide Wavelength  ************
510 DISP "Enter ";L$;"(cm) (0 to Quit)";@ INPUT LAM
520 IF LAM=0 THEN 410
530 IF LAM>= LAMCF THEN DISP "Use ";L$;"<";LAMCF @ BEEP @ GOTO 510
540 X=LAM @ GOSUB 580
550 PR$=L$&"= "&VAL$ (LAM)&" cm,  Guide "&L$&"= "
560 PR$=PR$&VAL$ (FNRD(LAMG))&" cm" @ GOSUB 880 @ GOTO 510
570 !
580 ! ********** Sub to calc guide wavelength *****************
590 LAMG=1/SQR (K/X/X-Y) @ RETURN
610 !
620 ! ***********   Enter antenna Parameters   ***************
630 DISP "Enter interelement spacing (cm)";@ INPUT d
640 IF d<= 0 THEN DISP "Use spacing > 0" @ BEEP @ GOTO 630
650 DISP "Enter delay line differential length (cm)";@ INPUT S
660 IF S<0 THEN DISP "Use length >= 0" @ BEEP @ GOTO 650
665 IF PN#1 THEN PRINT "Spacing=";d;" cm,  Delay Diff Length=";S;" cm"
670 DISP "Enter dielectric const (default=1)";@ INPUT K$@ K=1
680 IF K$#"" THEN K=VAL (K$)
690 IF K<1 THEN DISP "Use K>=1" @ BEEP @ GOTO 670
710 PR$="Dielectric. Const="&VAL$ (K) @ GOSUB 880 @ K1=S*SQR (K)
720 !
730 DISP "Select phase shift,F, (w/o scan) and delay line:"
740 DISP "  1  0 deg, Waveguide"
750 DISP "  2 180 deg,   ''"
760 DISP "  3  0 deg, Coax"
770 DISP "  4 180 deg,Coax"
780 DISP "Enter #";@ INPUT R@ Y,FLAG=0
790 IF R<1 OR R>4 THEN DISP "Use # 1-4" @ BEEP @ GOTO 780
800 F=360*FP (R/2-.5)
805 PR$="F="&VAL$ (F)&"deg, "
810 IF R<3 THEN FLAG=1 @ DISP "Enter wide W/G dim.(cm)";ELSE 850
820 INPUT a@ IF a<= 0 THEN DISP "Use a>0" @ BEEP @ GOTO 810
830 LAMCF=2*a/SQR (K) @ PR$=PR$&"W/G delay line, a= "&VAL$ (a)&" cm, Cut-off "
840 PR$=PR$&L$&"="&VAL$ (FNRD(LAMCF))&" cm" @ Y=.25/a/a @ GOTO 880
845 !
850 LAMCF=INF  @ PR$=PR$&"  Coaxial delay line "
860 !
870 ! ****************** DISP/PRINT *****************
880 DISP PR$ @ IF PN#1 THEN PRINT PR$
890 RETURN
900 !
910 ! *************  Scan Calculations  *****************
920 DISP "Enter ";L$;"for ";CF$;"(cm)";@ INPUT R@ IF R<= 2*d THEN 940
930 DISP "Use ";L$;"<=";2*d;" to avoid 'supergain' condition" @ BEEP @ GOTO 920
940 IF R<LAMCF THEN 942
941 DISP "Use ";L$;"< ";LAMCF;"cm, the W/G cutoff freq." @ BEEP @ GOTO 920
942 IF PN#1 THEN PRINT L$;"(";CF$;")=";R;"cm"
945 i=1 @ LAM,LAMC=R @ DPH=0 @ TPH=F/360 @ GOSUB 1130 @ THS,THCM=THi
950 DISP "Enter scan angle at ";CF$;"(deg)(default:";THS;"for 0 phase shift)";
960 INPUT R$@ IF R$#"" THEN THS=VAL (R$)
970 IF ABS (THS)>90 THEN DISP "Use angle<=90 deg" @ BEEP @ GOTO 950
980 DPH=360*FP (d/LAMC*(SIN (THS)-SIN (THCM))) @ TPH=(DPH+F)/360
990 IF PN#1 THEN PRINT "Scan Angle at ";CF$;"=";FNRD(THS);"deg,"
1000 PR$="Interelement phase shift (w/o F)="&VAL$ (FNRD(DPH))&" deg"
1005 GOSUB 880
```

```
1010 !
1020 DISP @ DISP "Scan Calc Menu: select next operation"
1030 DISP "1  New Center freq or scan angle"
1040 DISP "2  Beam-Normal ";L$;"s"
1050 DISP "3  Beam angles vs ";L$
1060 DISP "4  Calculations using antenna length"
1070 DISP "5  Main Menu"
1080 DISP @ DISP "Enter #";@ INPUT R
1090 IF R<1 OR R>5 THEN DISP "Use # 1-5" @ BEEP @ GOTO 1080
1100 ON R GOTO 920,1180,1310,1410,410
1110 !
1120 ! ************  sub to calc beam angle(s)  *****************
1130 SFLAG=0 @ X=LAM @ IF X>=LAMCF THEN SFLAG=1 @ RETURN
1135 GOSUB 580 @ n90=S/LAMG+TPH-d/LAM
1140 XX=FP (n90)+d/LAM @ IF n90<= 0 THEN XX=XX+1
1150 LL=LAM/d*(XX-i) @ IF LL<-1 THEN SFLAG=1 ELSE THi=ASN (LL)
1160 RETURN
1170 !
1180 ! ****************   Calc Beam-Normal Wavelengths  *****************
1190 IF S THEN 1230
1200 IF TPH MOD 360 THEN PR$="No" ELSE PR$="Any"
1210 PR$=PR$&L$&" yields a normal beam" @ GOTO 1005
1220 !
1230 X=FP (TPH) @ A=1 @ IF X THEN A=.5+SGN (X)/2-X
1240 FOR I=1 TO 5
1250 X=1/SQR (A*A/K1/K1+Y)
1260  PR$="Beam normal "&L$&"="&VAL$ (FNRD(X))&" cm" @ GOSUB 880 @ A=A+1
1270 NEXT I
1280 DISP "More ";L$;"s";@ INPUT R$@ R$=UPC$ (R$) @ IF R$="Y" THEN 1240
1290 IF R$#"N" THEN DISP "Y or N?" @ BEEP @ GOTO 1280 ELSE 1020
1300 !
1310 ! ************   Calc beam angles vs wavelength  *****************
1320 DISP "Enter ";L$;"(cm) (0 to quit) ";@ INPUT LAM
1330 DR$=L$&"="&VAL$ (LAM)&" cm, "
1340 IF LAM THEN i=1 @ GOSUB 1130 ELSE 1020
1350 IF SFLAG THEN PR$=DR$&"no visible beam!" @ GOSUB 880 @ GOTO 1320
1360 PR$=DR$&" angle="&VAL$ (FNRD(THi))&" deg"
1370 GOSUB 880 @ i=i+1 @ GOSUB 1150
1380 IF SFLAG THEN 1320 ELSE 1360
1390 !
1400 ! **************  Length Dependent calculations  **************
1410 DISP "Enter antenna length (cm)";@ INPUT L
1420 IF L<2*d THEN DISP "Use length >=";2*d @ BEEP @ GOTO 1410
1430 DISP "Enter Spreading factor for illum fn (>=1)";@ INPUT SF
1440 IF SF<1 THEN DISP "Use SF>=1" @ BEEP @ GOTO 1430
1450 BWN=50.76*SF*LAMC/L
1460 IF PN#1 THEN PRINT "Antenna length=";L;" cm, Spreading factor=";SF
1470 PR$="Normal beamwidth at "&CF$&"="&VAL$ (FNRD(BWN))&" deg"
1475 GOSUB 880
1480 !
1490 DISP @ DISP "Length dependent Menu, select next operation"
1500 DISP "1  Scan sensitivity and bandwidth"
1510 DISP "2  Scan coverage calcs"
1520 DISP "3  Scan Calc Menu"
1530 DISP "4  Main Menu"
```

```
1540 DISP "Enter #";@ INPUT R
1550 IF R<1 OR R>4 THEN DISP "Use # 1-4" @ BEEP @ GOTO 1540
1560 ON R GOTO 1580,1660,1020,410
1570 !
1580 ! ********   calc scan sens at angle THS, LAMC *****************
1590 LAM=LAMC/(1+LAMC/29979) @ i=1 @ GOSUB 1130 @ SENS=THi-THS @ BWMZ=INF
1600 IF SENS THEN BWMZ=BWN/2/SENS/COS (THS)
1620 PR$="Scan Sens="&VAL$ (SENS)&" deg/MHz, Bandwidth="&VAL$ (FNRD(BWMZ))
1630 PR$=PR$&" MHz" @ GOTO 1475
1640 !
1650 ! *****************  Scan Coverage Calcs  ******************
1660 DISP "Enter max scan angle(deg)";@ INPUT THM@ SA=SIN (ABS (THM))
1670 SB=2*SIN (BWN/2) @ dmax=LAMC/(1+SA+SB) @ NB=INT (SA/SB)+1
1680 PR$="Max Scan Angle="&VAL$ (THM)&" deg, max spacing="&VAL$ (FNRD(dmax))
1690 PR$=PR$&" cm, # of beams="&VAL$ (NB) @ GOTO 1475
1700 !
1710 ! *****************  initialization, printer selection  **********
1720 PAGESIZE 24 @ CLEAR @ DEG @ PRINTER IS 1
1730 DISP TAB (10);T$ @ DISP
1740 DISP "Select Printer:" @ DISP
1750 DISP "1  CRT"
1760 DISP "2  HP-82905B--HP-IB"
1770 DISP "3  HP-82905B--Centronics"
1780 DISP "4  MX-100"
1790 DISP "Enter #";@ INPUT PN
1800 IF PN<1 OR PN>4 THEN DISP "Use # 1-4 " @ BEEP @ GOTO 1790
1805 IF PN#1 THEN DISP "Make sure Top-Of-Form is set properly, Enter
date(no',')";@ INPUT DA$
1810 ON PN GOTO 1890,1830,1840,1860
1820 !
1830 PRINTER IS 301 @ GOTO 1850
1840 PRINTER IS 701
1850 PRINT CHR$ (27)&"&l1L" @ GOTO 1880
1860 PRINTER IS 701 @ PRINT CHR$ (27);"N";CHR$ (8)
1880 PRINT TAB (15);T$ @ PRINT TAB (25);DA$
1890 DISP @ DISP "Answer questions with 'Y' or 'N', [END LINE]" @ DISP
1900 DISP "All angles are in degrees, lengths in cm." @ DISP
1910 RETURN
```

Program ANTSYN

```
10 ! Pgm ANTSYN, rev 2/13/84
20 ! Translated from Program 1-4 of Radar Calc. , W. Skillman
30 !
40 ! Improvements vs Radar Calc.:
50 !    Data Correction
60 !
70 ! Important Variables:
80 !    EA...Synthesized pattern gain (voltage)
90 !    ES...Sampled (desired) pattern    "
100 !   GSL..mainlobe-to-sidelobe ratio(dB), Taylor weighting
110 !   L....Antenna length, same units as LAM
120 !   LAM..Operating wavelength
125 !   n....index of pattern points
130 !   r....mainlobe to sidelobe voltage ratio, Taylor wts
140 !   R4...n-bar, Taylor wts
150 !   SIG..sigma, Taylor weighting
160 !   TH...Angle from antenna normal,-90 to 90 deg
170 !   X....Normalized aperture dimension,-.5 to .5
180 !
190 DIM ES(101),EA(101),PR$[80],T$[35],AM(301),PH(301)
200 T$="**Antenna Pattern Synthesis**" @ L$="Wavelength "
210 !
220 ! *********  Defined Functions  ************************
230 DEF FNVDB(X) = 20*LGT (MAX (EPS ,ABS (X))) ! Voltage to dB
240 DEF FNIVDB(X) = 10^(X/20) !              dB to Voltage
250 DEF FNPDB(X) = FNVDB(X)/2 ! Power to dB
260 DEF FNRD(X) = INT (10000*X+.5)/10000 ! Round to 4 decimals
270 !
280 GOSUB 1190 ! init, printer selection
290 GOSUB 420 ! enter antenna length, wavelength
300 GOSUB 520 ! enter desired pattern
310 !
320 DISP @ DISP "Main Menu: select next operation" @ DISP
330 DISP "1  New length/";L$
340 DISP "2  New desired pattern"
350 DISP "3  Correct data"
360 DISP "4  Calc Synthesized pattern"
370 DISP "5  Calc Aperture function"
375 DISP "6  Disc Menu"
380 DISP @ DISP "Enter # ";@ INPUT R@ IF R=6 THEN CHAIN "Autost"
390 IF R<1 OR R>6 THEN DISP "Use # 1-6" @ BEEP @ GOTO 380
400 ON R GOTO 290,300,730,840,1000
410 !
420 ! ***********   Enter antenna length, wavelength ***********
430 DISP "Use same units for antenna length and ";L$
440 DISP "Enter length, ";L$;@ INPUT L,LAM
450 IF L<= 0 OR LAM<= 0 THEN DISP "Use #s > 0" @ BEEP @ GOTO 440
460 IF PN#1 THEN PRINT "Antenna length=";L;",   ";L$;LAM
465 RETURN
470 !
```

```
480 ! ************* Disp/Print **************************
490 DISP PR$ @ IF PN#1 THEN PRINT PR$
500 RETURN
510 !
520 ! ********* Enter desired antenna pattern *************
530 DISP "Select Desired Pattern" @ DISP
540 DISP "1  Taylor Weighted"
550 DISP "2  User-Defined"
560 DISP @ DISP "Enter #";@ INPUT R@ IF R=1 THEN 1400
570 IF R<1 OR R>2 THEN DISP "Use 1 or 2" @ BEEP @ GOTO 560
590 DISP "Enter approx max angle";@ INPUT THM
600 IF ABS (THM)>90 THEN DISP "Use angle <= 90 deg" @ BEEP @ GOTO 590
610 IF PN#1 THEN PRINT "    n         Angle (deg)       Gain (dB)"
620 N=INT (L/LAM*SIN (THM))+1 @ I=1 @ SUME,SUME2=0
630 IF N*LAM/L>= 1 THEN N=N-1 @ GOTO 630 ELSE NMAX=N
640 M=N*LAM/L @ IF M<= -1 THEN DISP "Last sample" @ RETURN
650  TH=FNRD(ASN (M)) @ DISP "Enter gain (dB WRT peak), for n=";
660  DISP N;", angle=";TH;" deg,[EL] to quit";@ INPUT G$
670  IF G$="" THEN NMIN=N+1 @ RETURN ELSE G=VAL (G$)
680  IF G>0 THEN DISP "Use gain < 0 dB" @ BEEP @ GOTO 650
690  X,ES(I)=FNIVDB(G)/SQR (COS (TH)) @ SUME=SUME+X @ SUME2=SUME2+X*X
700  IF PN#1 THEN PRINT TAB (5);N,TH,G
710 I=I+1 @ N=N-1 @ GOTO 640
720 !
725 ! *************** Data Correction ****************************
730 DISP "Enter n and gain (dB)";@ INPUT N,G
750 IF N>NMAX OR N<NMIN THEN DISP "Use ";NMIN;"<=n<=";NMAX @ BEEP @ GOTO 730
760 IF G>0 THEN DISP "Use Gain <=0 " @ BEEP @ GOTO 730
770 TH=FNRD(ASN (N*LAM/L)) @ ESN=FNIVDB(G)/SQR (COS (TH)) @ I=NMAX-N+1
780 ESL=ES(I) @ SUME=SUME-ESL+ESN @ SUME2=SUME2-ESL*ESL+ESN*ESN
790 ES(I)=ESN @ IF PN#1 THEN PRINT TAB (5);N,TH,G
800 DISP "More corrections";@ INPUT R$@ R$=UPC$ (R$) @ IF R$="Y" THEN 730
820 IF R$#"N" THEN DISP "Y or N?" @ BEEP @ GOTO 800 ELSE 320
830 !
840 ! ************** Sweep Synthesized Pattern ****************
850 DISP "Enter max angle, increment, min angle (deg) ";@ INPUT THT,DTH,THB
860 IF THB>= THT THEN DISP "Use max angle > min angle" @ BEEP @ GOTO 850
870 IF DTH=0 THEN DISP "Use increment > 0" @ BEEP @ GOTO 850
880 PR$=" Angle (deg)      Gain (dB)" @ GOSUB 490
890 J=0 @ FOR TH=THT TO THB STEP -ABS (DTH) @ K,SSUM=0 @ J=J+1
900   FOR I=NMAX TO NMIN STEP -1 @ K=K+1
910     T=L/LAM*SIN (TH) @ X=T-I @ GOSUB 960 @ SSUM=SINC*ES(K)+SSUM
920   NEXT I @ X,EA(J)=SQR (COS (TH)/SUME2)*SSUM
930   PR$="    "&VAL$ (TH)&"              "&VAL$ (FNRD(FNVDB(X))) @ GOSUB 480
940 NEXT TH @ GOTO 320
950 !
960 ! ************** SINC FUNCTION ********************
970 IF FP (X) THEN SINC=SIN (180*X)/PI /X @ RETURN
980 SINC=1 @ IF X THEN SINC=0
990 RETURN
1000 !
```

```
1010 ! ************ Sweep  Aperture Function  ****************
1020 DISP "Enter increment of x (aperture -.5<=x<=.5)";@ INPUT DX
1040 IF DX>.5 OR DX<= 0 THEN DISP "Use 0< increment <=.5" @ BEEP @ GOTO 1020
1050 PR$="   X         Amp (dB)     Phase (deg)" @ GOSUB 490
1060 PR$="   0          0             0 " @ GOSUB 490
1070 J=1 @ FOR X=DX TO .5 STEP DX @ GOSUB 1090 @ J=J+1 @ NEXT X @ GOTO 320
1080 !
1090 ! ****** Aperture weight using Goertzel DFT algorithm  *******
1100 I2,I3=0 @ PHF=360*X @ CP=COS (PHF) @ FF=2*CP
1110 FOR I=NMAX-NMIN+1 TO 1 STEP -1
1120    I1=ES(I)-I3+FF*I2 @ I3=I2 @ I2=I1
1130 NEXT I @ AI=I1-I3*CP @ AQ=-(I3*SIN (PHF))
1140 Y,AM(J)=FNPDB((AI*AI+AQ*AQ)/SUME^2) @ PH=(ATN2 (AQ,AI)+PHF*NMAX) MOD 360
1150 IF PH<0 THEN PH=PH+360 @ GOTO 1150 ELSE PH(J)=PH
1160 PR$="   "&VAL$ (X)&"        "&VAL$ (FNRD(Y))&"
1170 PR$=PR$&VAL$ (FNRD(PH)) @ GOTO 490
1180 !
1190 ! ************ INITIALIZATION &  PRINTER SELECTION  ***********
1200 PAGESIZE 24 @ CLEAR @ DEG @ PRINTER IS 1
1210 DISP TAB (10);T$ @ DISP
1220 DISP "Select Printer:" @ DISP
1230 DISP "1   CRT"
1240 DISP "2   HP-82905B--HP-IB"
1250 DISP "3   HP-82905B--Centronics"
1260 DISP "4   MX-100"
1270 DISP "Enter #";@ INPUT PN
1280 IF PN<1 OR PN>4 THEN DISP "Use # 1-4 " @ BEEP @ GOTO 1270
1285 IF PN#1 THEN DISP "Make sure Top-Of-Form is set properly, Enter
date(no',')";@ INPUT DA$
1290 ON PN GOTO 1370,1310,1320,1340
1300 !
1310 PRINTER IS 301 @ GOTO 1330
1320 PRINTER IS 701
1330 PRINT CHR$ (27)&"&l1L" @ GOTO 1360
1340 PRINTER IS 701 @ PRINT CHR$ (27);"N";CHR$ (8)
1360 PRINT TAB (15);T$ @ PRINT TAB (25);DA$
1370 RETURN
1380 !
1390 ! *************** Calc Taylor Coefficients  ***********
1400 IF PN#1 THEN PRINT "Taylor Weighted Pattern"
1410 GOSUB 1670 ! Enter GSL, Calc. n-bar minimum
1420 GOSUB 1610 ! Enter n-bar
1430 GOSUB 1780 ! Calc. sigma
1440 PR$="Sigma="&VAL$ (SIG) @ GOSUB 480 @ R2=1 @ FSUM,F2SUM=.5
1450 JU,JL=R4 @ NMAX=R4-1 @ ES(JU)=1 @ NMIN=-NMAX
1460 FOR M=1 TO NMAX
1470    M2=M*M @ X=.5 @ H8=M2/SIG/SIG
1480    FOR N=1 TO NMAX
1490      X=X*(1-H8/(ASQ+(N-.5)*(N-.5)))
1500      IF N#M THEN X=X/(1-M2/N/N)
1510    NEXT N @ X=R2*X @ R2=-R2
1520    JU=JU+1 @ JL=JL-1 @ ES(JU),ES(JL)=X
1530    FSUM=FSUM+X @ F2SUM=F2SUM+X*X
```

```
1540    PR$="F"&VAL$ (M)&"="&VAL$ (X) @ GOSUB 480
1550 NEXT M @ SUME=2*FSUM @ SUME2=2*F2SUM
1560 X=r/SQR (2)´ @ GOSUB 1810
1570 X=SIG*SQR (5.0968*(ASQ-X))
1580 PR$="Widening factor= "&VAL$ (X) @ GOTO 480
1590 !
1600 ! ******************** SUB A OF PGM 5-8, ENTER n-bar  *******
1610 DISP "n-bar =";@ INPUT R4
1620 IF R4>100 THEN DISP "Use n-bar <= 100 " @ BEEP @ GOTO 1610
1630 IF PN#1 THEN PRINT "n-bar =";R4
1635 RETURN
1640 !
1650 ! ************* SUB B OF PGM 5-8, ENTER GSL IN DB  **********
1660 ! Calc. n-bar minimum
1670 DISP "Enter GSL (dB)";@ INPUT X@ GSL=ABS (X)
1680 IF PN#1 THEN PRINT "GSL =";-GSL;" dB"
1690 X,r=10^(GSL/20) @ R4=1 @ SIG=0
1700 GOSUB 1810 @ ASQ=X ! Calc A^2
1710 T=SIG @ R4=R4+1
1720    GOSUB 1780 ! Calc sigma
1730 IF SIG>T THEN 1710
1740 ! Exit when sigma begins to decrease
1750 PR$="n-bar min ="&VAL$ (R4-1) @ GOTO 480
1760 !
1770 ! ********** SUB 107 OF PGM 5-8, CALC SIGMA  *******
1780 SIG=R4/SQR (ASQ+(R4-.5)*(R4-.5)) @ RETURN
1790 !
1800 ! ********** SUB B' OF PGM 5-8, CALC A^2 **********
1810 X=SQR (X*X-1)+X @ X=LOG (X)/PI  @ X=X*X @ RETURN
```

Chapter 2 Propagation

Program ANGLES

```
10 ! Pgm ANGLES, Rev. 2/05/84
20 ! Translation of Pgm 2-1 of Radar Calc...by W. Skillman
30 !
40 ! Improvements vs. Radar Calc.:
50 !    Calc of min bounce delay for known target elev case  (D')
60 !    Eq. 2-1.11 numerator factors, G,GT added
70 !
80 ! Important variables (angles in degrees)
90 !   (Note: elevation angles are + UP, depression angles + DOWN)
100 !    AL....alpha, grazing angle at bounce point
110 !    BRT...central angle, radar to target(Eq. 2-1.2)
120 !    DA$...date
130 !    DP....depression angle of ground point
140 !    DR....bounce range delay (nmi)
150 !    DRM...measured DR (nmi)
160 !    EHOR..elev of radar horizon
170 !    EP....elev of ground bounce point
180 !    ET.... "    "  target
190 !    ETT... "    "  radar at target
200 !    ETTP..90 + ETT (see Eq. 2-1.1)
210 !    FPMI..6076.115 = feet/nmi
220 !    G.....1+HR/RE
230 !    GT....1+HT/RE
240 !    HR....radar alt, nmi (HRFT in feet)
250 !    HT....target alt, nmi (HTFT in feet)
260 !    PN....printer code
270 !    RE....4588nmi = 4/3rds Earth radius
280 !    RG....range of ground point (nmi)
290 !    RHOR..radar horizon(nmi)
300 !    RR....range to bounce point (nmi)
310 !    RT.... "    "  target (nmi)
320 !    RTM...meas. target range (nmi)
330 !    RTT...range from bounce pt to target (nmi)
340 !
350 DIM PR$[80],T$[30]
360 RE=4588 @ FPMI=6076.115 @ T$="**TARGET ANGLES/RANGES**"
370 DR$=CHR$ (7)&"R" ! "DELTA" R, for display only
380 !
390 ! ***********   Defined Functions ********************
400 DEF FNRD(X) = INT (1000*X+.5)/1000 ! Round to 3 decimals
410 DEF FNc(a,b,ANG) = SQR (a*a+b*b-2*a*b*COS (ANG)) ! cos law, side opp ANG
420 DEF FNANG(a,b,c) = ACS ((a*a+b*b-c*c)/2/a/b) ! cos law, ang opp side c
430 DEF FNFEP(EP) = FRR*(1+(SIN (ET-EP)-SIN (2*AL))/SIN (2*AL-ET+EP))-DR
440 !
450 GOSUB 1590 ! init, select printer
460 GOSUB 600 ! enter radar altitude, calc hor. range, Elevation
470 !
```

```
480 ! **********   MAIN MENU   ****************************
490 PR$="" @ GOSUB 670 @ DISP "Main Menu: Select operation"
500 DISP "1  Known target range & alt, calc tgt elev, bounce el, delay"
510 DISP "2  Ground Point angles"
520 DISP "3  Known target elev, bounce delay,or elev,calc tgt R, alt"
530 DISP "4  Known target range and bounce delay, calc tgt altitude"
540 DISP "5  New Radar Altitude" @ DISP "6  Disc Menu"
550 DISP @ DISP "Enter #";@ INPUT R@ IF R=6 THEN CHAIN "Autost"
560 IF R<1 OR R>6 THEN DISP "Use # 1-6" @ BEEP @ GOTO 550
570 ON R GOSUB 700,930,1040,1350,600 @ GOTO 490
580 !
590 ! ***********   ENTER Radar alt, calc horiz range, elev  **********
600 DISP "Enter radar altitude (ft)";@ INPUT HRFT
610 IF HRFT<= 0 THEN DISP "Use alt>0" @ BEEP @ GOTO 600
620 HR=HRFT/FPMI @ G=1+HR/RE @ EHOR=-ACS (1/G) @ RHOR=SQR (HR*RE*(1+G))
630 PR$="Radar alt="&VAL$ (HRFT)&"ft, Elev of horiz="
640 PR$=PR$&VAL$ (FNRD(EHOR))&"deg,   Horiz range="&VAL$ (FNRD(RHOR))&"nmi"
650 !
660 ! *************************  DISPLAY/PRINT  *******************
670 DISP PR$ @ IF PN#1 THEN PRINT PR$
680 RETURN
690 !
700 ! ***   SUB TO CALC TGT ELEV, BOUNCE EL, DELAY, GIVEN TGT ALT, R**
710 DISP "Enter target altitude (ft)([EL] to quit)";@ INPUT R$@ R$=UPC$ (R$)
720 IF R$="" THEN RETURN ELSE HTFT=VAL (R$) @ HT=HTFT/FPMI
730 IF HT<= 0 THEN DISP "Use Alt >0" @ BEEP @ GOTO 710
740 DISP "Enter target range (nmi)(0 TO QUIT)";@ INPUT RT@ IF RT=0 THEN 710
750 IF RT<0 THEN DISP "Use range>=0" @ BEEP @ GOTO 740
760 IF PN#1 THEN PRINT "Target Alt (ft)=";HTFT;", Range (nmi)=";RT
770 ETTP=FNANG(RT,HT+RE,HR+RE) ! 90+ETT
780 BRT=ASN (RT/(HR+RE)*SIN (ETTP))
790 ET=90-BRT-ETTP @ IF ET>EHOR THEN 810
800 IF RT>RHOR THEN PR$="Below Horizon" @ GOSUB 670 @ GOTO 740
810 PR$="Target Elev ="&VAL$ (ET)&"deg" @ GOSUB 670
820 GOSUB 870 ! Elev and range delay of bounce
830 PR$="Bounce pt elev ="&VAL$ (FNRD(EP))&"deg"&", Bounce delay="
840 PR$=PR$&VAL$ (DR)&"nmi" @ GOSUB 670 @ GOTO 740
850 !
860 ! ******* SUB TO CALC BOUNCE EL, DELAY  FROM TGT H, R  *********
870 AL=ATN ((HT+HR)/SQR (RT*RT-(HR-HT)^2)) @ GT=1+HT/RE
880 CAL=COS (AL) @ SAL=SIN (AL) @ EPT=-ACS (CAL/GT)
890  EP=-ACS (CAL/G) @ F=2*AL+BRT+EPT+EP @ XT=GT*SIN (EPT) @ XR=G*SIN (EP)
900  FF=2+SAL/XT+SAL/XR @ DAL=F/FF @ AL=AL-DAL
910 IF ABS (DAL)>.0001 THEN 880 ELSE DR=-(RE*(XR+XT+2*SAL))-RT @ RETURN
920 !
930 ! **** sub to calc graz, dep at ground point  ***********
940 DISP "Enter Range (nmi) to ground point (0 to Quit)";@ INPUT RG
950 IF RG=0 THEN RETURN
960 IF RG<= HR THEN DISP "Use range>";HR @ BEEP @ GOTO 940
970 PR$="Ground point range="&VAL$ (RG)&"nmi, "
980 AL=FNANG(RG,RE,RE+HR)-90
990 IF AL<0 THEN PR$=PR$&"Below horizon" @ GOTO 1020
1000 DP=ACS (COS (AL)/G) @ PR$=PR$&"Graz ang="
```

```
1010 PR$=PR$&VAL$ (FNRD(AL))&"deg, Dep ang="&VAL$ (FNRD(DP))&"deg"
1020 GOSUB 670 @ GOTO 940
1030 !
1040 ! ********  CALC TGT R,H FROM TGT EL, BOUNCE RANGE OR EL   ******
1050 DISP "Enter target elev (deg) ([EL] to Quit)";@ INPUT ET$
1060 IF ET$="" THEN RETURN ELSE ET=VAL (ET$)
1070 IF PN#1 THEN PRINT "Target elev (deg) = ";ET
1080 DRMIN=0 @ IF ET<EHOR THEN 1110
1090 RT=2000 @ HT=FNc(RT,HR+RE,90+ET)-RE
1100 BRT=ASN (RT*SIN (90+ET)/(RE+HT)) @ GOSUB 870 @ DRMIN=DR
1110 DISP "Minimum bounce delay (nmi)=";DRMIN
1120 DISP "Enter tgt bounce ";DR$;"(nmi) or Elev(deg) ([EL] to Quit)";
1130 INPUT R$@ IF R$="" THEN 1050 ELSE EP,DR=VAL (R$)
1140 IF DR<0 THEN 1310
1150 IF DR>2*HR THEN DISP "Use bounce ";DR$;"<";2*HR @ BEEP @ GOTO 1120
1160 IF DR<DRMIN THEN DISP "Use bounce ";DR$;">";DRMIN @ BEEP @ GOTO 1120
1170 IF PN#1 THEN PRINT "Bounce delay =";DR;"nmi"
1180 EP=-9 @ DEP=.0001 ! Newton-Raphson root finding
1190 EPP=EP @ GOSUB 1520 @ Y=FNFEP(EP)
1200  EPP=EP+DEP @ GOSUB 1520 @ Z=FNFEP(EPP)
1210  IF Y=Z OR Y=0 THEN 1280 !  Exit if delay too short for solution
1220  X=DEP/(1-Z/Y) @ EP=EP+X @ IF EP>EHOR THEN 1280 ! "
1230 IF ABS (X)>DEP THEN 1190 ! iterate until error small
1240 EPP=EP @ GOSUB 1520 @ RT=FRR*SIN (2*AL)/SIN (2*AL-ET+EP)
1250 HT=FNc(RT,HR+RE,90+ET)-RE
1260 PR$="Tgt range="&VAL$ (FNRD(RT))&"nmi, Alt ="
1270 PR$=PR$&VAL$ (INT (FPMI*HT+.5))&"ft" @ GOSUB 670 @ GOTO 1120
1280 BEEP @ DISP "Bounce delay too small, try again!" @ GOTO 1120
1290 !
1300 ! ********  CALC TGT R,H FROM TGT EL & BOUNCE EL
1310 IF EP>EHOR THEN DISP "Use elev<=";EHOR @ BEEP @ GOTO 1120
1320 IF PN#1 THEN PRINT "Bounce elev =";EP;"deg"
1330 GOTO 1240
1340 !
1350 ! ********  SUB TO CALC TGT H, GIVEN TGT R, BOUNCE DELAY  *********
1360 DISP "Enter target range (nmi) (0 to quit)";@ INPUT RTM
1370 IF RTM=0 THEN RETURN
1380 IF RTM<0 THEN DISP "Use R>=0" @ BEEP @ GOTO 1360
1390 DISP "Enter bounce delay (nmi) (0 to quit)";@ INPUT DRM
1400 IF DRM=0 THEN 1360
1410 IF DRM<0 THEN DISP "Use delay >=0" @ BEEP @ GOTO 1390
1420 IF DRM>2*HR THEN DISP "Use bounce ";DR$;" < ";2*HR @ BEEP @ GOTO 1390
1430 EP=-9 @ DEP=.0001 ! Newton-Raphson root finding
1440 EPP=EP @ GOSUB 1560 @ Y=FEP2
1450  EPP=EP+DEP @ GOSUB 1560 @ Z=FEP2 @ IF Y=Z THEN 1470 ELSE X=DEP/(1-Z/Y)
1460 EP=EP+X @ IF ABS (X)>DEP THEN 1440
1470 HT=FNc(RE,RTT,90+AL)-RE @ HTFT=HT*FPMI
1480 PR$="Range="&VAL$ (RTM)&"nmi, Bounce delay="&VAL$ (DRM)&"nmi"
1490 PR$=PR$&", Alt="&VAL$ (INT (HTFT+.5))&"ft" @ GOSUB 670 @ GOTO 1390
1500 !
1510 ! ***********  Range: radar to bounce point  ******************
1520  CEP=COS (EPP) @ CC=CEP*G @ IF CC>1 THEN AL,FRR=0 @ RETURN
1530  AL=ACS (CC) @ FRR=RE*SIN (-EPP-AL)/CEP @ RETURN
```

```
1540 !
1550 ! *********  SUB TO CALC EP2, RANGE DIFFERENCE  **************
1560   GOSUB 1520 @ RR=FRR @ RTT=RTM+DRM-RR
1570   RT=FNc(RR,RTT,180-2*AL) @ FEP2=RT-RTM @ RETURN
1580 !
1590 ! ***********  INITIALIZATION &  PRINTER SELECTION  ***************
1600 PAGESIZE 24 @ CLEAR @ DEG @ PRINTER IS 1
1610 DISP TAB (10);T$ @ DISP
1620 DISP "Select Printer:" @ DISP
1630 DISP "1   CRT"
1640 DISP "2   HP-82905B--HP-IB"
1650 DISP "3   HP-82905B--Centronics"
1660 DISP "4   MX-100"
1670 DISP "Enter #";@ INPUT PN
1680 IF PN<1 OR PN>4 THEN DISP "Use # 1-4 " @ BEEP @ GOTO 1670
1685 IF PN#1 THEN DISP "Make sure Top-Of-Form is set properly, Enter
date(no',')";@ INPUT DA$
1690 ON PN GOTO 1770,1710,1720,1740
1700 !
1710 PRINTER IS 301 @ GOTO 1730
1720 PRINTER IS 701
1730 PRINT CHR$ (27)&"&11L" @ GOTO 1760
1740 PRINTER IS 701 @ PRINT CHR$ (27);"N";CHR$ (8)
1760 PRINT TAB (15);T$ @ PRINT TAB (25);DA$
1770 ! DISP @ DISP "Answer questions with 'Y' or 'N', [END LINE]" @ DISP
1780 RETURN
```

Program TRACE

```
10 ! Pgm TRACE,2/12/84 calculates radar range-height, angle chart for
20 !    exponential atmosphere, trans. from Prog. 2-2 of Radar Calc...
30 ! Written by W. Skillman
40 !
50 ! Improvements vs. Radar Calc...
60 !    Atmospheric attn. curve calculated from wavelength, stored each 10kft
70 !       Straight line interpolation used
80 !    Fixed segment length (lnmi currently)
90 !    Attenuation output whenever dB equals preset value;i.e. .25,.5,.75 etc.
100 !
110 ! Important Variables: (all angles in degrees)
120 !    AM.......Atmospheric data array:RHO(gm./cu.m.),Pressure(mbar),
125 !       T(deg-K), every 10kft, starting at 0
130 !    AT.....dB values for dB atten curve output
140 !    ATT....Atmospheric loss rate(dB/nmi), at 10kft intervals
150 !    BETA...Central angle from ray end to radar
160 !    DH.....altitude spacing of outputs (ft)
170 !    DR.....range increment (nmi)
180 !    EL.....ray elevation from local horizontal, + up
190 !    FLAG3..=1 if H exceeded HMAX, =0 otherwise
200 !    GAMi...atten rate of Oxy, waterl, water2
210 !    H......ray altitude (nmi)
220 !    HH.....output alt (ft)
230 !    HMAX...max ray altitude (ft)
240 !    HR.....radar altitude (ft)
250 !    LFLAG..=1, Calc atmos loss during trace, =0, don't
260 !    LOSS...cumulative one-way atmos. attn along ray  (dB)
270 !    NPTS...no. of traces to run
280 !    NS.....surface refractivity (313 default)
290 !    PN.....printer code
300 !    RR.....radius from center of Earth to ray end (nmi)
310 !    RS.....cumulative slant range along ray path    "
320 !    RSP....output slant range (nmi)
330 !    R12....NS/1E6
340 !    SQT... sq root of T
350 !    T......Temp (deg-K)
360 !    T2.....square of T
370 !    WL.....wavelength (cm)
380 !    WL2....    "      squared
390 !    X......range in rect. coord (nmi)
400 !    Y......alt. in rect. coord. (ft)
410 DIM PR$[80],NS$[25],RA$[25],ELV(50),RH$[25],AT(28),B$[30],C$[30]
420 DIM AM(3,11),ATT(11)
430 DEF FNn(H) = 1+R12*R18^H ! Refractivity, n
440 NS$="Surface Refractivity" @ RH$="Radar Altitude (ft)"
450 ! dB values for atmos. loss output
460 DATA .25,.5,.75,1,1.5,2,3,4,5,6,7,8,9,10,12,15,20,25,30,35,40,45,50,60
470 DATA 70,80,90,100
480 FPMI=6076.115 @ T$="** RAY TRACE  ***" @ WL$=" Wavelength "
490 FOR I=1 TO 28 @ READ AT(I)@ NEXT I ! Loss values for dB output
500 !
```

```
510 ! Atmospheric Data: RHO, P and T at 10kft increments starting at 0
520 DATA 7.5,1013.25,288.16,2.23,696.943,268.357,.452,466.003,240.574
530 DATA .05233,301.484,228.809,.0040128,188.23,216.66,.000846,116.641,216.66
540 DATA .00055,72.312,216.66,.000621,44.851,216.66,.000804,27.831,216.66
550 DATA .000733,17.377,216.66,.00044,11.053,216.66
560 FOR I=1 TO 11
570   FOR J=1 TO 3 @ READ AM(J,I)@ NEXT J
580 NEXT I
590 GOSUB 2060 ! init, printer sel.
600 DISP "Select operation:"
610 DISP "1  Ray Trace"
620 DISP "2  Chart Elevations"
630 DISP "Enter #";@ INPUT R
640 IF R<1 OR R>2 THEN DISP "Use 1 or 2" @ BEEP @ GOTO 630
650 IF R=2 THEN SFLAG=1 @ GOTO 1600
660 GOSUB 1820 ! Select/deselect atmos loss calc.,enter wvln, calc attn
670 SFLAG=0
680 GOSUB 1480 ! Enter radar altitude
690 GOSUB 1420 ! Enter Surface Refractivity
700 GOSUB 1540 ! enter alt. increments and range/alt limits
710 GOSUB 1350 ! enter elev. angles & do traces
720 !
730 ! ********** MAIN MENU ********************
740 IF SFLAG THEN 660 ELSE DISP "Select next operation" @ DISP
750 DISP "1  New radar altitude"
760 DISP "2  New Surface refractivity"
770 DISP "3  New alt. increments and/or range/alt limits"
780 DISP "4  More elevation angles"
790 IF LFLAG THEN DISP "5 ";WL$
800 DISP "6  Select/deselect atmos loss calc"
810 DISP "7  Chart Elevations" @ DISP "8  Disc Menu"
820 DISP @ DISP "Enter #";@ INPUT R @ IF R=8 THEN CHAIN "Autost"
830 IF R<1 OR R>8 THEN DISP "Use # 1-8" @ BEEP @ GOTO 820
840 ON R GOSUB 1480,1420,1540,1350,1870,1820,1600 @ GOTO 740
850 !
860 ! ********** sub for single ray trace **********************
870 PR$="" @ GOSUB 1110 @ RSP,ATTN,X,BETA,RS,LL,LOSS=0 @ J,DR=1 ! delta R=1nmi
880 H=HR/FPMI @ RR=H+3440
890 EL,EL0=ELV(I) @ Y,HH=HR
900 PR$="    Range    Altitude    Elevation    X        Y        Loss"
910 GOSUB 1110
920 PR$="    (nmi)      (ft)       (deg)      (nmi)    (ft)       (dB)"
930 GOSUB 1110 @ nold=FNn(H) @ R3=HR/DH @ GOSUB 1300
940 IF H<= 0 THEN 1070 ! end trace if ray intersects the ground
950 HOLD=H @ GOSUB 1150 ! delta H test, calc atmos loss
960 IF FLAG3 THEN RETURN ! End trace if beyond max alt.
970 DISP H;RS;EL;LL ! monitor progress
980 IF RS>RMAX THEN RETURN ! End trace if beyond max range
990 RS=RS+DR @ LOSS=LOSS+DR*LL
1000 RRN=SQR (RR*RR+DR*DR+2*RR*DR*SIN (EL)) ! new radius
1010 H=H+RRN-RR @ RR=RRN ! update alt, radius
1020 DBETA=ASN (DR*COS (EL)/RRN) ! increment of central angle
```

```
1030 EL=EL+DBETA @ BETA=BETA+DBETA @ nnew=FNn(H)
1040 EL=SGN (EL)*ACS (nold/nnew*COS (EL)) @ nold=nnew @ GOTO 940 ! new elev.
1050 !
1060 !   **************   Wrap up at Earth Intersection *************
1070 RSP=RS+DR*(H/(HOLD-H)) @ HH=0 @ ATTN=LOSS @ GOSUB 1280
1080 PR$="Grazing angle ="&VAL$ (-EL-BETA)&"deg"
1090 !
1100 ! *****************   SUB TO DISPLAY/PRINT   ********************
1110 DISP PR$ @ IF PN#1 THEN PRINT PR$
1120 RETURN
1130 !
1140 ! ***********  sub to calc atmos loss, max alt test, alt output **
1150 HFT=H*FPMI @ HF=HFT/10000 @ IH=INT (HF)+1
1160 IF HFT>HMAX THEN FLAG3=1 ELSE FLAG3=0
1170 IF LFLAG=0 THEN 1230
1180 IF HFT>100000 THEN LL=ATT(11) ! Assume const loss rate beyond table
1190 LL=ATT(IH)-FP (HF)*(ATT(IH)-ATT(IH+1))
1210 DL=LOSS-AT(J) @ IF DL<0 THEN 1230 ELSE DDR=DL/LL
1220 RSP=RS-DDR @ HH=HFT-DDR*FPMI*SIN (EL) @ ATTN=AT(J) @ GOSUB 1280 @ J=J+1
1230 R4=R3 @ R3=HFT/DH @ X=INT (R3) @ Y=INT (R4)
1240 IF X=Y THEN RETURN ! return if alt boundary not crossed
1260 H6=MAX (X,Y) @ HH=H6*DH @ ATTN=LOSS
1270 RSP=RS+DR*(H6-R3)/(R3-R4) ! Interpolate range at crossing
1275 IF RSP=0 THEN RETURN
1280 X=RSP*COS (EL0) @ Y=RSP*FPMI*SIN (EL0)+HR ! RECT. coord of crossing
1290 F1: IMAGE 3X,DDD.DDD,3X,3DC3D,5X,MDD.DD,3X,DDD.D,3X,M3DC3D,3X,DD.DD
1300 DISP USING F1 ; RSP,HH,EL,X,Y,ATTN
1310 IF PN#1 THEN PRINT USING F1 ; RSP,HH,EL,X,Y,ATTN
1320 RETURN
1330 !
1340 ! *********   sub to enter elev. angles for trace(s)   *************
1350 NPTS=0
1360 DISP "Enter elev. angle (deg)of ray ([END LINE] to start trace(s))";
1370 INPUT EL$@ IF EL$="" THEN 1390 ELSE NPTS=NPTS+1
1380 ELV(NPTS)=VAL (EL$) @ IF NPTS<50 THEN 1360
1390 FOR I=1 TO NPTS @ GOSUB 870 @ NEXT I @ RETURN
1400 !
1410 ! ***********   SUB TO ENTER & CALC SURFACE REFRACTIVITY CONSTS   *****
1420 DISP "Enter ";NS$;" (313 default)";@ INPUT NSS$
1430 IF NSS$="" THEN NS=313 ELSE NS=VAL (NSS$)
1440 IF PN#1 THEN PRINT "  Surface Refractivity, Ns = ";NS
1450 R18=(1-7.32*1.00559258^NS/NS)^1.852 @ R12=NS/1000000 @ RETURN
1460 !
1470 ! **********   sub to enter radar altitude   *******************
1480 DISP "Enter ";RH$;@ INPUT HR
1490 IF HR<0 THEN DISP "Use Alt.>=0" @ BEEP @ GOTO 1480
1500 IF PN#1 THEN PRINT RH$;" = ";HR
1510 RETURN
1520 !
```

```
1530 ! *********** sub to enter chart alt increments and limits  *******
1540 DISP "Enter altitude increment for chart (ft)";@ INPUT DH
1550 IF DH<= 0 THEN DISP "Use # >0 " @ BEEP @ GOTO 1540
1560 DISP "Enter Max. Altitude (ft) and Max. Range (nmi)";@ INPUT HMAX,RMAX
1570 IF HMAX<= 0 OR RMAX<= 0 THEN DISP "Use # > 0" @ BEEP @ GOTO 1560
1580 RETURN
1590 !
1600 ! ***************** Calc. Chart Elevations  *******************
1610 DISP "Input Vertical Expansion Factor, K";@ INPUT K
1620 IF K<= 0 THEN DISP "Use K>0" @ BEEP @ GOTO 1610
1630 IF PN#1 THEN PRINT "  K=";K
1640 DISP "Enter space elev. (deg) ([END LINE] to quit)";@ INPUT THS$
1650 IF THS$="" THEN 740 ELSE THS=VAL (THS$) @ THC=ATN (K*TAN (THS))
1660 PR$="Space Elev="&THS$&"deg,    Chart Elev ="&VAL$ (THC)&"deg"
1670 GOSUB 1110 @ GOTO 1640
1680 !
1690 ! ********** SUB TO CALC OXYGEN ATTEN  ****************
1700 DV1=P*(T^(-.75))/795 @ DV2=DV1*795/287
1710 X=1/(1/WL2/DV1+DV1)+1/((2+1/WL)^2/DV2+DV2)+1/((2-1/WL)^2/DV2+DV2)
1720 GAM1=53.35*P*X/WL2/T2 @ RETURN
1730 !
1740 ! **************** SUB TO CALC WATER ATTEN  ************
1750 DV3=P*(1+.0046*RHO)/653/SQT
1760 X=1/((1/WL-.741)^2/DV3+DV3)+1/((1/WL+.741)^2/DV3+DV3)
1770 GAM2=86544*EXP (-(644/T))*RHO*X/WL2/T2/SQT @ RETURN
1780 !
1790 ! ******** SUB TO CALC WATER ATTEN ABOVE 10KFT  *************
1800 GAM3=DV3*RHO*27.13/WL2/T @ RETURN
1810 !
1820 ! ******* sub to select atmos loss calc or not ***********
1830 LFLAG=1 @ DISP "Do you wish to calc. atmos loss during trace (Y/N)";
1840 INPUT R$@ R$=UPC$ (R$) @ IF R$="N" THEN LFLAG=0 @ RETURN
1850 IF R$#"Y" THEN DISP "Y or N?" @ BEEP @ GOTO 1830
1860 !
1870 ! *********** ENTER WAVELENGTH, CALC ATTN CURVE AND STORE  *****
1880 DISP "Enter";WL$;"(cm) (>0.6 cm)";@ INPUT WL
1890 IF WL<.6 THEN DISP WL$;"must be greater than 0.6 cm" @ BEEP @ GOTO 1880
1900 WL2=WL*WL @ PR$="  "&WL$&"(cm)="&VAL$ (WL) @ GOSUB 1110
1910 PR$="  Altitude   Loss Rate    Oxy rate    water rate1   water rate2"
1920 GOSUB 1110
1930 PR$="    (ft)      (dB/nmi)    (dB/nmi)    (dB/nmi)      (dB/nmi)"
1940 GOSUB 1110
1950 F3: IMAGE 2X,3DC3D,3X,D.DDDE,3X,D.DDDE,3X,D.DDDE,3X, D.DDDE
1960 FOR I=1 TO 11
1970    RHO=AM(1,I) @ P=AM(2,I) @ T=AM(3,I)
1980    T2=T*T @ SQT=SQR (T) @ Y=(I-1)*10000
1990    GOSUB 1700 @ GOSUB 1740 @ GOSUB 1800
2000    X,ATT(I)=GAM1+GAM2+GAM3
2010    DISP USING F3 ; Y,X,GAM1,GAM2,GAM3
2020    IF PN#1 THEN PRINT USING F3 ; Y,X,GAM1,GAM2,GAM3
2030 NEXT I @ RETURN
2040 !
2050 !
```

```
2060 ! ***********  INITIALIZATION &  PRINTER SELECTION  ***************
2070 PAGESIZE 24 @ CLEAR @ DEG @ PRINTER IS 1
2080 DISP TAB (10);T$ @ DISP
2100 DISP "Select Printer:" @ DISP
2110 DISP "1  CRT"
2120 DISP "2  HP-82905B--HP-IB"
2130 DISP "3  HP-82905B--Centronics"
2140 DISP "4  MX-100"
2150 DISP "Enter #";@ INPUT PN
2160 IF PN<1 OR PN>4 THEN DISP "Use # 1-4 " @ BEEP @ GOTO 2150
2165 IF PN#1 THEN DISP "Make sure Top-Of-Form is set properly, Enter
date(no',')";@ INPUT DA$
2170 ON PN GOTO 2250,2190,2200,2220
2180 !
2190 PRINTER IS 301 @ GOTO 2210
2200 PRINTER IS 701
2210 PRINT CHR$ (27)&"&l1L" @ GOTO 2240
2220 PRINTER IS 701 @ PRINT CHR$ (27);"N";CHR$ (8)
2240 PRINT TAB (20);T$ @ PRINT TAB (25);DA$
2250 RETURN
```

Program COVER

```
10 ! Pgm COVER, Rev 2/05/84
20 ! Translated from Prog. 2-3 of Radar Calc.... By W. Skillman
30 !
40 ! Improvements vs. Radar Calc:
50 !    Constant far sidelobe antenna gain
60 !    Selectable range units
70 !
80 ! Important variables (angles in degrees)
90 !    BW....1/2 power beamwidth at THP=THNO
100 !   LDB...range dependent loss in dB
110 !   RP....Performance range (ex, range at which Pd=.35)
120 !   TH....angle from peak of beam
130 !   THNO..angle of normal of antenna to reference
140 !   THP...Angle of peak of beam WRT reference
150 !
160 DIM A$[30],B$[35],C$[40],D$[30],T$[45],E$[25],PR$[80],QR$[50],G$[25]
170 !
180 A$="1/2 Power beamwidth (deg)"
190 B$="angle of beam peak WRT ref (deg)"
200 C$="angle of normal to ant. WRT ref (deg) " @ G$="User-Defined Beamshape"
210 D$="Performance range at beam peak" @ E$="gain WRT pk (dB)"
220 T$="**Off-Boresight Range Performance**" @ F$="Use angle <= 90deg"
230 DATA (km),(nmi),(mi),(ft)
240 FOR I=1 TO 4 @ READ RI$(I)@ NEXT I @ GOTO 340
250 !
260 ! ******************* Defined Functions *******************
270 DEF FNRD(X) = INT (X*100+.5)/100 ! Round to 2 decimals
280 DEF FNPDB(X) = 10*LGT (X) ! Power ratio to dB
290 ! FNGAIN(TH) ! Gaussian beamshape, TH=ang from peak,X=Pwr Gain
300 X=(SIN (TH-THNO)-SIN (THP-THNO))/SIN (BW)
310 IF ABS (X)>XMAX THEN X=GSL ELSE X=COS (TH-THNO)*(2^(-(4*X*X)))
320 RETURN ! Power WRT peak
330 !
340 GOSUB 1020 ! init & printer selection
350 GOSUB 480 ! Enter antenna parameters
360 GOSUB 670 ! enter performance range
370 GOSUB 800 ! calc. off-boresight performance
380 !
390 ! ************ Main Menu *******************
400 PR$="" @ GOSUB 990 @ DISP "Select next operation "
410 DISP "1  New antenna parameters"
420 DISP "2  New Performance range"
430 DISP "3  Off-boresight performance" @ DISP "4  Disc Menu"
440 DISP @ DISP "Enter #";@ INPUT R@ IF R=4 THEN CHAIN "Autost"
450 IF R<1 OR R>4 THEN DISP "Use # 1-4" @ BEEP @ GOTO 440
460 ON R GOSUB 480,670,800 @ GOTO 390
470 !
```

```
480 ! **********  sub to enter antenna parameters  ***********
490 DISP @ DISP "Enter ";A$;" (0 for ";G$;")";@ INPUT BW
500 IF BW<0 THEN DISP "Use # >=0" @ BEEP @ GOTO 490
510 UDFLAG=0 @ IF BW THEN 540 ELSE UDFLAG=1
520 IF PN#1 THEN PRINT G$
530 GOTO 580
540 IF PN#1 THEN PRINT A$;"=";BW
550 DISP "Enter ";B$;@ INPUT THP
560 IF ABS (THP)>90 THEN DISP F$ @ BEEP @ GOTO 550
570 IF PN#1 THEN PRINT B$;"=";THP
580 DISP "Enter ";C$;@ INPUT THNO
590 IF ABS (THNO)>90 THEN DISP F$ @ BEEP @ GOTO 580
600 IF PN#1 THEN PRINT C$;"=";THNO
610 IF UDFLAG THEN RETURN
620 DISP "Enter far sidelobe ";E$;@ INPUT R@ GSLDB=ABS (R)
630 XMAX=SQR (GSLDB/40/LGT (2)) @ GSL=10^(-(GSLDB/10))
640 IF PN#1 THEN PRINT "Far ";E$;"= -";GSLDB
650 RETURN
660 !
670 ! **********   SUB TO ENTER PERFORMANCE RANGE  **************
680 DISP @ DISP "Select Range Units: " @ DISP
690 DISP "1  kilometers (km)"
700 DISP "2  nautical miles (nmi)"
710 DISP "3  statute miles (mi)"
720 DISP "4  feet (ft)"
730 DISP "Enter #";@ INPUT RI
740 IF RI<1 OR RI>4 THEN DISP "Use # 1-4" @ BEEP @ GOTO 730
750 DISP @ DISP "Enter ";D$;RI$(RI);@ INPUT RP
760 IF RP<= 0 THEN DISP "Use Range>0" @ BEEP @ GOTO 750
770 IF PN#1 THEN PRINT D$;"=";RP;RI$(RI)
780 RETURN
790 !
800 ! *********  CALC OFF-BORESIGHT RANGE PERF.  *************
810 DISP "Enter target angle from ref (deg)([END LINE] to Quit)";@ INPUT R$
820 IF R$="" THEN RETURN
830 TH=VAL (R$) @ IF ABS (TH)>90 THEN DISP F$ @ BEEP @ GOTO 810
840 QR$="Angle = "&VAL$ (FNRD(TH))&"deg, "
850 IF UDFLAG=0 THEN GOSUB 290 @ DG=X @ GOTO 890
860 DISP "Enter ";E$;", angle WRT normal = ";TH-THNO;@ INPUT DGDB
870 IF DGDB>0 THEN DISP "Use gain <0(dB)" @ BEEP @ GOTO 860
880 DG=10^(DGDB/10) @ QR$=QR$&E$&"="&VAL$ (DGDB)
890 R,RR=RP*SQR (DG)
900 PR$=QR$&", Range="&VAL$ (FNRD(R))&RI$(RI)
910 DISP "Range Dependent loss iteration: enter 0 dB to quit"
920 DISP "Enter 1-way range-dependent loss in dB for ";R;RI$(RI);@ INPUT LDB
930  IF LDB=0 THEN GOSUB 990 @ GOTO 810 ELSE LDB=ABS (LDB)
940  R=RR*(10^(-(LDB/20)))
950  PR$=QR$&"Loss (dB)= -"&VAL$ (LDB)&",    Range="&VAL$ (FNRD(R))&RI$(RI)
960 GOTO 920
970 !
```

```
980  ! **********   DISPLAY/PRINT   **************************
990   DISP PR$ @ IF PN#1 THEN PRINT PR$
1000 RETURN
1010 !
1020 ! **********   INITIALIZATION &  PRINTER SELECTION  ***************
1030 PAGESIZE 24 @ CLEAR @ DEG @ PRINTER IS 1
1040 DISP TAB (10);T$ @ DISP
1050 DISP "Select Printer:" @ DISP
1060 DISP "1   CRT"
1070 DISP "2   HP-82905B--HP-IB"
1080 DISP "3   HP-82905B--Centronics"
1090 DISP "4   MX-100"
1100 DISP "Enter #";@ INPUT PN
1110 IF PN<1 OR PN>4 THEN DISP "Use # 1-4 " @ BEEP @ GOTO 1100
1115 IF PN#1 THEN DISP "Make sure Top-Of-Form is set properly, Enter
date(no',')";@ INPUT DA$
1120 ON PN GOTO 1210,1140,1150,1170
1130 !
1140 PRINTER IS 301 @ GOTO 1160
1150 PRINTER IS 701
1160 PRINT CHR$ (27)&"&11L" @ GOTO 1200
1170 PRINTER IS 701 @ PRINT CHR$ (27);"N";CHR$ (8)
1200 PRINT TAB (15);T$ @ PRINT TAB (25);DA$
1210 RETURN
```

Program MONOERR

```
10  ! Pgm MONOERR , Rev 2/05/84
20  ! translated from Pgm 2-4, Radar Calc.., by W. Skillman
25  !
30  ! Improvements vs. Radar Calc.
40  !   Small angle approximations removed
50  !   Automatic grazing angle sweep
60  !
70  ! Important variables: (angles input in degrees, radians used in pgm)
80  !   A$.....annotation array
90  !     1...rel. diel const          10...range (nmi)
100 !     2...surf. cond. (mhos/m)     11...beam elev (deg)
110 !     3...rms surf ht dev (ft)     12...polarization
120 !     4...rms surf slope dev (deg) 13...Fresnel refl coef
130 !     5...elev beamwidth (deg)     14...Specular pwr
140 !     6...veg. fract absorp        15...Diffuse pwr
150 !     7...wavelength (cm)          16...rms elev error (mr)
160 !     8...radar ht (ft)            17...Nosediving
170 !     9...target ht (ft)
180 !
190 !   AL.....grazing angle
200 !   AL2....grazing angle from tgt(deg)
210 !   ELT....elev of tgt (rad)
220 !   FLAG1..1=clutter patch within beam width, 0=otherwise
230 !   FLAG4..0=Vertical, 1=Horizontal polarization
240 !   flag5..0=diffuse, 1=specular calc.
250 !   HR.....radar height (ft)
260 !   HT.....target   "      "
270 !   Ii ....diffuse power density
280 !   IIS....specular power
290 !   LAM....wavelength (ft)
300 !   P......parameter array (corresponds to A$ array, 1-11)
310 !   RFT....target range (ft)
320 !   SIG....rms elev error (mr)
330 !   SUMI...sum of diffuse power
340 !
350 DATA relative dielectric constant,surface conductivity(mhos/m)
360 DATA rms surface height deviation (feet),rms surface slope deviation(deg)
370 DATA elevation beamwidth (deg),Veg fractional absorption,wavelength (cm)
380 DATA radar altitude (ft)
390 DATA target altitude (ft),target range (nmi),elev. of beam peak (deg)
400 DATA polarization,Fresnel reflection coef.=
410 DATA Specular power=,Diffuse power=,rms elev error(mr)=,Nosediving
420 !
430 DIM A$(17)[40],P(14),PR$[60],T$[45]
440 !
450 ! ************* Defined Functions ***********************
460 DEF FNRD(X) = INT (X*1000+.5)/1000 ! Round to 3 decimals
470 !
```

```
480 T$="***Monopulse Multipath Tracking Error***" @ FPMI=6076.115
490 FOR I=1 TO 17 @ READ A$(I)@ NEXT I @ P4=PI *PI /4 @ R2D=180/PI
500 GOSUB 1370 ! Init, printer selection
510 GOSUB 600 ! enter initial set of parameters
520 !
530 PR$="" @ GOSUB 880 @ DISP "Main Menu: select next operation"
540 DISP "1  New Parameters"
550 DISP "2  Calc Multipath error" @ DISP "3  Disc Menu"
560 DISP @ DISP "Enter #";@ INPUT R @ IF R=3 THEN CHAIN "Autost"
570 IF R<1 OR R>3 THEN DISP "Use # 1-3" @ BEEP @ GOTO 560
580 ON R GOSUB 1200,680 @ GOTO 530
590 !
600 ! ********** sub to enter parameters  ******************
610 DISP "Parameters may be reviewed and revised following initial entry"
620 FOR I=1 TO 11 @ GOSUB 1330 @ NEXT I
630 DISP "Enter 0 for vertical, 1 for horizontal ";A$(12);@ INPUT R
640 IF R<0 OR R>1 THEN DISP "Use 0 or 1 " @ BEEP @ GOTO 630
650 IF R THEN PR$="Horizontal " ELSE PR$="Vertical "
660 PR$=PR$&A$(12) @ FLAG4=R @ GOTO 880
670 !
680 ! ********** SUB TO CALC MULTIPATH ERROR  ***************
690 RFT=P(10)*FPMI @ HR=P(8) @ HT=P(9) @ RAD ! nmi to feet
700 AL=ATN ((HR+HT)/SQR (RFT*RFT-(HT-HR)^2)) ! Grazing angle of specular pt.
710 ELT=ASN ((HT-HR)/RFT) @ EN=ATN (HR/RFT/COS (ELT)) ! tgt el, nadir dep.
720 LAM=P(7)/30.48 @ BZ=P(4)/R2D @ GOSUB 910 ! CALC u
730 FLAG1=0 @ IF u<5.78 THEN FLAG1=1
740 FLAG5=1 @ GOSUB 940 @ DISP A$(13);RHOZ @ BEEP 20,150 @ BEEP 15,320
750 DISP " Press [EL] for this value, or enter your #, press [EL]";
760 INPUT R$@ IF R$#"" THEN X,RHOZ=VAL (R$) @ IIS=IIS*X*X/RHOZ2
770 PR$=A$(13)&VAL$ (FNRD(RHOZ)) @ GOSUB 880
780 PR$=A$(14)&VAL$ (IIS) @ GOSUB 880
790 X=RHOZ*SQR (RHOS2) @ IF X>.5 AND FLAG1 THEN 1180
800 Ii,SUMI,FLAG5=0 @ DAL=P(5)/R2D/10 @ AL=EN+DAL/2
810 IOLD=Ii @ GOSUB 940 @ SUMI=SUMI+Ii ! Diffuse part (143)
820    PR$="Grazing angle="&VAL$ (FNRD(AL*R2D))&"deg, "&A$(15)&VAL$ (Ii)
830 GOSUB 880 @ AL=AL+DAL @ IF Ii>IOLD THEN 810 ! next loop if pwr increase
840 IF Ii>SUMI/1000 THEN 810 ! convergence test, increase "1000", more accy
850 SIG=500*P(5)*P(6)*SQR ((SUMI+IIS)/2)/R2D @ PR$=A$(16)&VAL$ (FNRD(SIG))
860 !
870 ! ************** DISPLAY/PRINT ********************
880 DISP PR$ @ IF PN#1 THEN PRINT PR$
890 RETURN
900 !
910 ! ********** SUB TO CALC u  ********************
920 u=3.77*(AL*R2D+P(11))/P(5) @ RETURN
930 !
940 ! ************ SUB TO CALC POWER COMPONENT (154) ************
950 AL2=ATN (HT/(RFT*COS (ELT)-HR/TAN (AL)))
960 ALAV=(AL+AL2)/2 @ a=AL/2/BZ @ c=AL2/2/BZ @ ap=MIN (a,c) @ cp=MAX (a,c)
970 SIGHP=P(3) @ IF ap<.25 THEN SIGHP=SIGHP*(4*ap) .2
980 ALC=LAM/4/PI /SIGHP @ b=ALC/2/BZ
990 RHOS2=EXP (-(ap*ap/b/b)) @ SAL=SIN (ALAV)
1000 IF b<1+ap THEN Z=8*SQR (1-RHOS2)/(1/ap+4+3*ap-b*b/3/ap-b) @ GOTO 1020
```

```
1010 Z=24/(2/ap/ap+9/ap+12+5*ap)
1020 XX=AL2*AL2/ALC/ALC @ YY=MIN (XX,500) ! Prevent underflow in Fd2 calc
1030 ZZ=AL*AL/ALC/ALC @ ZZ=MIN (ZZ,500) !      ditto
1040 Fd2=SQR ((1-EXP (-ZZ))*(1-EXP (-YY)))
1050 ETAd=ALAV/BZ/SQR (PI )/(AL-HR/RFT)*EXP (-((ap-cp)^2))*Fd2*Z/2
1060 RE=P(1)-COS (ALAV)^2 @ IM=60*P(2)*LAM @ AM=SQR (SQR (RE*RE+IM*IM))
1070 ANG=ATN2 (IM,RE)/2 @ RE=AM*COS (ANG) @ IMN,IMD=AM*SIN (ANG)
1080 IF FLAG4 THEN 1100 ELSE REA=SAL*P(1) @ IMA=IM*SAL
1090 RN=REA-RE @ RD=REA+RE @ IMN=IMA-IMN @ IMD=IMA+IMD @ GOTO 1110
1095 !
1100 RN=SAL-RE @ RD=SAL+RE
1110 RHOZ2=(RN*RN+IMN*IMN)/(RD*RD+IMD*IMD) @ RHOZ=SQR (RHOZ2)
1120 GOSUB 920 @ Pu=P4-u*u @ X=Pu*SIN (u)-2*u*COS (u)
1130 DEL2=36*X*X/Pu^4 @ Ii=DEL2*RHOZ2
1140 IF FLAG5 THEN IIS=Ii*RHOS2 ELSE Ii=Ii*ETAd*DAL
1150 RETURN
1160 !
1170 ! ********** sub for indication of nosediving ******************
1180 PR$=A$(17)&", RHOZ*RHOS="&VAL$ (X) @ GOSUB 880 @ GOTO 800
1190 !
1200 ! *************** SUB TO CHANGE PARAMETERS ***************
1210 FOR I=1 TO 11
1220    DISP I;TAB (7);P(I);TAB (23);"= ";A$(I)
1230 NEXT I @ DISP I;
1240 IF FLAG4 THEN PR$="Horizontal " ELSE PR$="Vertical "
1250 DISP PR$;A$(12)
1260 DISP @ DISP "Enter # of parameter to change (0 to quit)";
1270 INPUT I@ IF I=0 THEN RETURN
1280 IF I<0 OR I>12 THEN DISP "Use # 1-12 " @ BEEP @ GOTO 1270
1290 IF I<12 THEN GOSUB 1330 ELSE GOSUB 630
1300 GOTO 1260
1310 !
1320 ! *********** sub to enter parameters ********************
1330 DISP "Enter ";A$(I);@ INPUT P(I)
1340 IF PN#1 THEN PRINT TAB (7);P(I);TAB (23);"= ";A$(I)
1350 RETURN
1360 !
1370 ! *********** INITIALIZATION & PRINTER SELECTION ***************
1380 PAGESIZE 24 @ CLEAR @ PRINTER IS 1
1390 DISP TAB (10);T$ @ DISP
1400 DISP "Select Printer:" @ DISP
1410 DISP "1   CRT"
1420 DISP "2   HP-82905B--HP-IB"
1430 DISP "3   HP-82905B--Centronics"
1440 DISP "4   MX-100"
1450 DISP "Enter #";@ INPUT PN
1460 IF PN<1 OR PN>4 THEN DISP "Use # 1-4 " @ BEEP @ GOTO 1450
1465 IF PN#1 THEN DISP "Make sure Top-Of-Form is set properly, Enter
date(no',')";@ INPUT DA$
1470 ON PN GOTO 1550,1490,1500,1520
1480 !
```

```
1490 PRINTER IS 301 @ GOTO 1510
1500 PRINTER IS 701
1510 PRINT CHR$ (27)&"&l1L" @ GOTO 1540
1520 PRINTER IS 701 @ PRINT CHR$ (27);"N";CHR$ (8)
1540 PRINT TAB (15);T$ @ PRINT TAB (25);DA$
1550 RETURN
```

Chapter 3 Detection Probability

Program DETPROB

```
10  ! Pgm DETPROB, Rev 2/27/84
20  ! Written  by W. SKILLMAN, Translated from
30  ! Pgms 3-1 to 3-5 of "Radar Calculations using the TI-59 ..."
40  !
50  ! Improvements vs. Radar Calc....:
60  !    Exact solution for Yb (option, Pgm 3-1)
70  !    Exact solution for Swerling Case 1,N>1 (eq. 3-1.15), and
80  !       for Case 3,N>2 (eq. 3-1.19)
90  !    Auto load of single look Marcum curve
100 !       (D0(N) calc. from Pgm 3-4)
110 !    Keyboard calc of Error Function (ERF), Factorial
120 !    CRT or Plotter plot, dump from CRT to printer
130 !
140 ! IMPORTANT VARIABLES:
150 !  B1-B5...CONSTANTS FOR Q(X)
160 !  C1-C4...CONSTANTS FOR INVERSE Q FN
170 !  D11.....THRESHOLD IN DB
180 !  DZN.....Do(N), S/N FOR PD=.99, non-fluct tgt
190 !  ERF.....ERF(X)
200 !  FL......FLUCTUATION MODEL NO.
210 !     1=ONE-DOMINANT (SWERLING 3 OR 4)    4=GENERALIZED  EXPONENTIAL
220 !     2=RAYLEIGH (SWERLING 1 OR 2)        5=CHI-SQUARE(Single look only)
230 !     3=LOG-NORMAL
240 !  F3......FLAG, SWERLING 1
250 !  H7......MARCUM'S FALSE ALARM NO., n
260 !  I...... INCOMPLETE GAMMA FUNCTION
270 !  KCOR....CORRELATION WINDOW FOR TGT RPT
280 !  KCS.....DUO-DEGREES OF FREEDOM FOR CHI-SQUARE DIST
290 !  KGE.....GENERAL EXPONENTIAL PARAMETER
300 !  M.......# OF DETECTS RQD FOR TGT RPT, OUT OF N LOOKS
310 !  MNFLG...NON-FLUCT CURVE FLAG: 0=NOT LOADED, 1=LOADED
320 !  NFM.....FLUCTUATION SPEED
330 !     1=NON-FLUCTUATING         3=LOOK-TO-LOOK
340 !     2=SCAN-TO-SCAN            4=PULSE-TO-PULSE
350 !  NI......1/2 # OF POINTS IN GAUSS-LEGENDRE INTEG.
360 !  NINT....# OF PULSES (VARIATES) INTEGRATED
370 !  NL......# OF LOOKS PER DWELL
380 !  PD......PROB OF DET
390 !  PFA.....PROB OF FALSE ALARM, PER CELL
400 !  PFR.....PROB OF FALSE REPORT (M/N)
410 !  Q.......Q(X)
420 !  RZ......Range at which S=N, Marcum's Idealized Range
430 !  R3......AVG S/N / COLLAPSING RATIO
440 !  R9......INCOMPLETE GAMMA FN TEST VALUE
450 !  R10.....COLLAPSING RATIO,RHO
460 !  R11.....Yb, THRESHOLD
470 !  R12.....NINT TIMES COLLAPSING RATIO
480 !  R13.....RHO FOR LOG-NORMAL MODEL
490 !  SN......SIGNAL-TO-NOISE (AVERAGE, BEFORE COLLAPSING)
500 !  S12.....SQR(R12)
```

```
510 !  SNFLG...TYPE OF S/N ENTRY:
520 !      1=NUMERIC    2=DECIBEL     3=R/Ro    4=Range (user-defined units)
530 !  SQRHO...SQR(2*LN(RHO)), LOG-NORMAL MODEL
540 !  WI......GAUSS-LEGENDRE WEIGHTS
550 !  X99.....S/N FOR Pd=.99
560 !  ZI......GAUSS-LEGENDRE  ARGUMENTS
570 !
580 DIM L$[60],PR$[75],T$[35],G$[30],B$(5)[30],D$[30],FF(201),X$[25],TL$[40]
590 DIM TR$[30],LL$[30],X$(4)[25],UD$[80]
600 GOTO 1400
610 !
620 ! ********** Subroutine & Defined Functions   *********************
630 !
640 ! Fn FNYB(nM,N) Calc. Threshold from Marcum's n(nM)and # of pulses,N
650 ! Exact solution using Incomplete Gamma Function (Option, PGM 3-1)
660 ! Newton-Raphson root-finding used
670 PFA=N*LOG (2)/nM @ X=SQR (-LGT (PFA)) @ S12=SQR (N) @ R8=0
680 Y=N-S12+2.3*X*(X+S12-1) @ IF N/nM>.0000000001 THEN R8=EXP (-PFA)
690 GOSUB 1250 @ IF R8 THEN X=R8-IGF ELSE X=IC-PFA
700 DEL=X/R4/N @ Y=Y+DEL*Y @ IF ABS (DEL)>.001 THEN 690 ELSE YB=Y @ RETURN
710 !
720 ! FNEX(X), Protected EXP(X)
730 IF X>1151 THEN EX=INF  @ RETURN
740 IF X<-1151 THEN EX=0 ELSE EX=EXP (X)
750 RETURN
760 !
770 ! FNCP(PP,MM,N): PROB OF MM-OUT-OF-N AT PROB=PP
780 IF N=1 THEN CP=PP @ RETURN
790 IF MM>N THEN DISP "Use M<=N" @ BEEP @ CP=0 @ RETURN
800 R15=1/PP-1 @ R7=1 @ R1=N-MM @ CP,PQ=PP^N
810 IF R1 THEN PQ=PQ*N/R7*R15 @ CP=CP+PQ ELSE RETURN
820 R7=R7+1 @ N=N-1 @ R1=R1-1 @ GOTO 810
830 !
840 ! FNLFAC(N): Natural LOG of factorial function,N>0 only
850 N=N+5 @ N2=N*N
860 U=(N-1)*LOG (N)-N+LOG (SQR (2*PI *N))+(1-(1/30+1/105/N2)/N2)/12/N
870 LFAC=U-LOG ((N-1)*(N-2)*(N-3)*(N-4)) @ N=N-5 @ RETURN
880 !
890 ! FNFACT(N): Factorial function, N>=0 only, (LN(N!) IF 253<N<1E10)
900 OFFLAG=0 @ X=1 @ IF N=0 THEN 960
910 IF FP (N)=0 AND N<15 THEN 950
920 GOSUB 840 @ X=LFAC @ IF X>1151 THEN OFFLAG=1 ELSE X=EXP (X)
930 GOTO 960 ! OFFLAG Shows that LN(N!) is returned
940 !
950 FOR I=2 TO N @ X=X*I @ NEXT I ! Exact, integer < 15
960 FACT=X @ RETURN
970 !
980 ! FNPFR(PFA): Prob. of False report from PFA, KCOR,M,NL
990 PP=PFA @ KK=KCOR @ GOSUB 1060 @ PP=PK @ N=NL @ MM=M @ GOSUB 770
1000 PFR=PFA*CP/PK @ RETURN
1010 !
1020 ! FNEE(A,X): Eval of (1-A^X) for A<1
1030 AA=-(X*LOG (A)) @ IF AA<.000001 THEN EE=AA*(1-AA/2) ELSE EE=1-A^X
1040 RETURN
1050 !
```

```
1060 ! FNPK(PP,KK): Eval of 1-(1-PP)^KK, PP<1
1070 X=PP*KK @ IF X<.000001 THEN PK=X*(1-(X-PP)/2) ELSE PK=1-(1-PP)^KK
1080 RETURN
1090 !
1100 ! FNQ(X): Q(X) M-L Pgm 14, sub B (requires FNZ, consts P,B1-B5)
1110 T=1/(ABS (X)*P+1) @ XQ=X @ GOSUB 1150
1120 Q=Z*T*(T*(T*(T*(T*B5-B4)+B3)-B2)+B1)
1130 ZX=Z @ Q=.5+SGN (XQ)*(Q-.5) @ X=XQ @ RETURN
1140 !
1150 ! Function Z(X)
1160 XZ=X @ X=X*X @ GOSUB 720 @ Z=1/SQR (MIN(EX,INF/6.3)*2*PI )
1165 X=XZ @ RETURN
1170 !
1180 ! Function ERF(X)
1190 XE=X @ X=X*SQR (2) @ GOSUB 1100 @ ERF=1-2*Q @ X=XE @ RETURN
1200 !
1210 ! FNIQ(X): Inverse Q(X), requires consts C1-C4
1220 Y=MIN (X,1-X) @ T=SQR (-LOG (Y*Y))
1230 IQ=(T-(C1+C2/T)/(C3*T+C4+1/T))*SGN (.5-X) @ RETURN
1240 !
1250 ! FNIGF(Y,N): Incomplete Gamma Function,outputs IGF,IC,R4
1260 R9=1.00001 @ R6,XX,X=1 @ R7=N !  IC=complement of IGF=1-IGF
1270 GOSUB 1290 @ IGF=1-IC @ RETURN
1280 !
1290 IF N=1 THEN X=Y @ GOSUB 720 @ IC=1/EX @ R4=Y*IC @ RETURN
1300 IF Y>= N THEN 1350
1310 R7=R7+1 @ X=X*Y/R7 @ GOSUB 1380 @ IF Z>= R9 THEN 1310 ! Termination test
1330 GOSUB 840 @ X=N*LOG (Y)-Y-LFAC @ GOSUB 720 @ R4=EX @ IC=1-XX*R4 @ RETURN
1340 !
1350 R7=R7-1 @ X=X*R7/Y @ GOSUB 1380 ! Sub for IGF, Y>=N
1360 IF Z>= R9 THEN 1350 ELSE GOSUB 1330 @ IC=(1-IC)*N/Y @ RETURN
1370 !
1380 XX=XX+X @ Z=XX/R6 @ R6=XX @ RETURN ! sub for IGF, Y<N
1390 !
1400 DEF FNRD(X) = INT (X*1000+.5)/1000 ! Round to 3 decimals
1410 !
1420 ! Function RE, round to 4 significant figures
1430 DEF FNRE(X) = 10^INT (LGT (X)-3)*INT (X/10^INT (LGT (X)-3)+.5)
1440 !
1450 DATA 4,.183435,.362684,.525532,.313707,.796666,.22381,.96029,.101229
1460 READ NI ! 1/2 # of points of integration
1470 FOR J=1 TO NI @ READ ZI(J)@ READ WI(J)@ NEXT J ! G-L Coefs & args.
1480 DATA .231642,1.330274,1.82126,1.78148,.356564,.319382
1490 DATA .27061,2.30753,.04481,.99229,***DETECTION PROBABILITY***
1500 READ P,B5,B4,B3,B2,B1,C1,C2,C3,C4,T$
1520 X$(1)="Signal-to-Noise" @ YL$="Prob. of Detection" @ ZZZ$="0"&CHR$ (210)
1530 X$(2)=X$(1)&" (dB)" @ X$(3)="R/Ro" @ DE$=CHR$(7)
1540 !
1550 GOSUB 6260 ! Init, Printer selection
1560 GOSUB 7040 ! Select Plotter, load binary GDUMP
1570 !
```

```
1580 ! ****************** MAIN MENU **********************
1590 DISP @ DISP "** Main Menu **  Select next operation" @ PR$="" @ GOSUB 5920
1600 DISP "1 Pd Calculation" @ DISP "2 Threshold"
1620 DISP "3 Incomplete Gamma Function" @ DISP "4 ERF,Q,Z Functions"
1640 DISP "5 Inverse Q Function" @ DISP "6 Marcum's Q(a,b)"
1660 DISP "7 PFR vs Pfa, M/N detector"
1670 DISP "8 Factorial "
1680 DISP "9 Disc menu"
1690 DISP @ DISP "Enter #";@ INPUT OPT
1700 IF OPT<1 OR OPT>9 THEN DISP "Use # 1-9" @ BEEP @ GOTO 1690
1710 ON OPT GOTO 1740,2640,3420,3160,5320,3530,3970,6120,6100
1720 !
1730 ! ***************** START Pd CALCULATION ***********
1740 GOSUB 3220 ! Select type of S/N entry
1750 GOSUB 3340 ! Enter N and RHO
1760 GOSUB 4250 ! Detection algorithm and Pfa
1770 GOSUB 4560 ! Fluctuation speed
1780 IF NFM>1 THEN GOSUB 4700 ! Fluctuation model
1782 DISP "Select Plot mode:" @ DISP @ DISP " 1 No Plot" @ DISP " 2 New Plot"
1784 IF PF>1 THEN DISP " 3 Overlay Plot"
1786 YB,DXS,II=0 @ DISP @ DISP "Enter #";@ INPUT PF
1788 IF PF<1 OR PF>3 THEN DISP "Use 1-3" @ BEEP @ GOTO 1786
1789 IF PF=3 THEN AP=1 @ SS=XL @ ON SNFLG GOTO 1880,2000,2110,2250
1790 AP=0 @ DX,LT,YT=1 @ ON SNFLG GOTO 1820,1940,2050,2180 ! S/N Sweep
1800 !
1810 ! ***************** S/N NUMERIC ***********
1820 DISP "Enter S/N (numeric) minimum (>=0)";@ INPUT SS@ XL=SS
1830 IF SS<0 THEN DISP "Use S/N>=0 " @ DISP @ BEEP @ GOTO 1820
1840 DISP "Enter S/N (numeric) maximum";@ INPUT XR
1850 IF SS>= XR THEN 1880
1860 DISP "Enter ";DE$;"S/N (numeric)";@ INPUT DX
1870 IF DX<= 0 THEN DISP "Use ";DE$;"S/N > 0" @ BEEP @ GOTO 1860
1875 IF (XR-XL)/DX>200 THEN DISP "Use # pts<201" @ BEEP @ GOTO 1820
1880 PR$="        S/N           Pd" @ GOSUB 5920
1890 F1: IMAGE 6X,D.DDDE,3X,D.3D
1895 IF SS=0 THEN PD=0 @ GOSUB 2620 @ GOTO 1910
1900 SN=SS @ GOSUB 2540
1910 SS=SS+DX @ IF SS<= XR THEN 1900 ELSE 2310
1920 !
1930 ! ************** S/N IN DECIBELS ***********
1940 DISP "Enter S/N (dB) Minimum";@ INPUT SS@ XL=SS
1950 DISP "Enter S/N (dB) Maximum ";@ INPUT XR
1960 IF SS>= XR THEN GOTO 1990
1970 DISP "Enter ";DE$;"S/N (dB)";@ INPUT DX
1980 IF DX<= 0 THEN DISP "Use ";DE$;" >0" @ BEEP @ GOTO 1970
1985 IF (XR-XL)/DX>200 THEN DISP "Use # pts<201" @ BEEP @ GOTO 1940
1990 F2: IMAGE 10X,3D.2D,11X,D.3D
2000 PR$="          S/N (dB)           Pd" @ GOSUB 5920
2010 SN=10^(SS/10) @ GOSUB 2540
2020 SS=SS+DX @ IF SS<= XR THEN 2010 ELSE 2310
2030 !
2040 ! ******************* S/N FROM R/Ro ***********
2050 DISP "Enter R/Ro minimum (>=0)";@ INPUT SS@ XL=SS
2060 IF SS<0 THEN DISP "Use R/Ro>=0" @ DISP @ BEEP @ GOTO 2050
```

```
2070 DISP "Enter R/Ro Maximum";@ INPUT XR
2080 IF SS>= XR THEN 2110
2090 DISP "Enter ";DE$;X$(3);@ INPUT DX
2100 IF DX<= 0 THEN DISP "Use ";DE$;" >0" @ BEEP @ GOTO 2090
2105 IF (XR-XL)/DX>200 THEN DISP "Use # pts<201" @ BEEP @ GOTO 2050
2110 PR$="           R/Ro           Pd" @ GOSUB 5920
2120 F3: IMAGE 8X,4D.DD,8X,D.3D
2130 IF SS=0 THEN PD=1 @ GOSUB 2620 @ GOTO 2150
2140 SN=SS^-4 @ GOSUB 2540
2150 SS=SS+DX @ IF SS<= XR THEN 2140 ELSE 2310
2160 !
2170 ! ********************** S/N FROM R  *****************
2180 DISP "Enter Ro (";RU$;")";@ INPUT RZ@ IF PN#1 THEN PRINT "Ro(";RU$;")=";RZ
2190 DISP "Enter R minimum (>=0)";@ INPUT SS@ XL=SS
2200 IF SS<0 THEN DISP "R>=0" @ DISP @ BEEP @ GOTO 2190
2210 DISP "Enter R Maximum";@ INPUT XR
2220 IF SS>= XR THEN 2250
2230 DISP "Enter ";DE$;"R";@ INPUT DX
2240 IF DX<= 0 THEN DISP "Use ";DE$;" >0" @ BEEP @ GOTO 2230
2245 IF (XR-XL)/DX>200 THEN DISP "Use # pts<201" @ BEEP @ GOTO 2190
2250 PR$="           R("&RU$&")          Pd" @ GOSUB 5920
2270 IF SS=0 THEN PD=1 @ GOSUB 2620 @ GOTO 2290
2280 SN=(RZ/SS)^4 @ GOSUB 2540
2290 SS=SS+DX @ IF SS<= XR THEN 2280
2310 ON PF GOSUB 1080,6560,6550 ! Plot mode
2330 !
2340 ! ************* Pd MENU  **************
2350 PR$="" @ GOSUB 5920
2360 DISP "Pd Menu: Select next operation " @ DISP
2370 DISP "1 More Points"
2380 IF NFM#1 THEN DISP "2 New Fluctuation Model"
2390 DISP "3 New Fluctuation speed"
2400 DISP "4 New Detection Algorithm and Pfa"
2410 DISP "5 New N and/or RHO"
2420 DISP "6 New S/N Entry Type" @ DISP "7 Main Menu"
2440 DISP "8 Program Menu " @ DISP
2450 DISP "Enter no. ";@ INPUT RR
2460 IF NFM=1 AND RR=2 THEN DISP "Don't use #2, only 1 model for Non-fluct
tgt." @ BEEP @ GOTO 2450
2470 IF RR<1 OR RR>8 THEN DISP "Use # 1-8" @ DISP @ BEEP @ GOTO 2450
2480 ON RR GOTO 1790,1780,1770,1760,1750,2510,1580,6100·
2490 !
2500 ! ******  NEW S/N ENTRY TYPE  ******
2510 GOSUB 3220 @ GOTO 2350
2520 !
2530 ! ********  Select Pd sub vs. fluct speed & model  *****
2540 R3=SN/R10
2550 IF FL=5 THEN GOSUB 3760 @ GOTO 2620 ! Chi-square model
2560 IF NL>1 THEN 5070 ! Branch for multilook calc, cont for single look
2570 IF NFM=2 AND FL>2 THEN 5070 ! Use M/N for L-N,GE
2580 IF NFM=1 THEN GOSUB 3060 ! Non-fluctuating
2590 IF NFM=2 THEN ON FL GOSUB 2840,2710 ! Scan-to-scan, SW 3 OR 1
2600 IF NFM=4 THEN ON FL GOSUB 2910,2790 ! Pulse-to-pulse, SW 4 OR 2
2610 ! NFM cannot=3 for single look/dwell
2620 II=II+1 @ FF(II)=PD @ ON SNFLG GOTO 5980,6010,6040,6040 ! disp/print Pd
```

```
2640 ! ********** SUB TO CALC. THRESHOLD FROM Pfa FROM KEYBOARD ***********
2650 PR$="" @ GOSUB 5920
2660 DISP "Enter Pfa, 0<Pfa<1 (Pfa=N*ln(2)/n) (O to Quit)";@ INPUT PFA
2670   IF PFA=0 THEN 1570
2680   IF PFA<0 THEN DISP "Use PFA>0 " @ DISP @ BEEP @ GOTO 2660
2690 GOSUB 3340 @ M,NL=1 @ GOSUB 4430 @ GOTO 2660 ! Enter N & RHO, calc Yb
2700 !
2710 ! ********************* SWERLING CASE 1  ********************
2720 X=R3*R12 @ R1=X+1 @ R2=X/R1
2730 X=LOG (R2)*(1-R12)-R11/R1 @ GOSUB 720 @ R5=EX ! EQ 3-1.16, X lg, approx
2740 IF R12=1 OR R3>2/(R11-R12) THEN 2770 ! Exit if N=1,Eq 3-1.14, Exact, or
2750 !  if S/N large, Eq. 3-1.16, approx, continue for Eq. 3-1.15, Exact
2760 Y=R11*R2 @ N=R12 @ GOSUB 1250 @ R5=R5*IGF @ Y=R11 @ GOSUB 1250 @ R5=R5+IC
2770 PD=R5 @ RETURN
2780 !
2790 ! ******************* SWERLING CASE 2 (417)  ********************
2800 R8=R3+1 @ IF R12<10 THEN Y=R11/R8 @ N=R12 @ GOSUB 1250 @ PD=1-IGF @ RETURN
2810 R8=R8*S12
2820 T=1/3/S12 @ X=.25 @ GOTO 3100
2830 !
2840 ! ********************** SWERLING CASE 3  ******************
2850 X3=R12*R3/2 @ X4=R11/(1+X3) ! NXbar/2
2860 X=X4 @ GOSUB 720 @ PD=(1+1/X3)^(R12-2)*(1+X4-(R12-2)/X3)/EX
2870 IF R12<3 THEN RETURN ! Eq.3-1.20 Exact for N=1,2
2880 Y=X4*X3 @ N=R12-1 @ GOSUB 1250 @ PD=PD*IGF ! Eq. 3-1.19, Exact for N>2
2890 Y=R11 @ GOSUB 1250 @ PD=IC+R4*N/(1+X3)+PD @ RETURN
2900 !
2910 ! *************** SWERLING CASE 4 (437)  ******************
2920 H8=R3/2 @ X,R8=1+H8
2930 IF R12<10 THEN 2980
2940 R5=2*R8*R8-1 @ R8=SQR (R5) @ X=2*X*X*X-1 ! N>=10
2950 T=X/3/S12/R8/R5 @ R8=R8*S12 @ X=T*T/2-.25 @ GOTO 3090
2960 !
2970 ! ******************** CASE 4, N<10  *******************
2980 R3=1 @ R1=R11/R8 @ Y=R1 @ H6,N=R12 @ GOSUB 1250 @ R5,R6=IGF ! Eq. 3-1.21
2990 FOR R2=R12 TO 1 STEP -1
3000   R3=R3*R2*H8/(R12-R2+1) @ H6=H6+1
3010   R6=R6-R4 @ R4=R4*R1/H6 @ R5=R5+R6*R3
3020 NEXT R2 @ PD=1-R5/R8^R12 @ RETURN
3030 !
3040 ! ***************** MARCUM CASE 5 (291)  **********************
3050 ! IF N=1, use Marcum's Q fn, else use Gram-Charlier series
3060 IF R12=1 THEN X=R3 @ GOSUB 3610 @ PD=Q @ RETURN
3070 R5=2*R3+1 @ R8=SQR (R5*R12) ! Gram-Charlier  series for N>1
3080 T=(R3+1/3)/R8/R5 @ X=R3+.25
3090 X=X/R5/R5
3100 R7=X/R12
3110 X=(R11-R12*(1+R3))/R8 @ R6=-T @ T=X*X
3130 Y=(3-T)*R7-R6*(R6*(15+T*(T-10))/2-X+1/X)
3140 GOSUB 1100 @ PD=Q-Y*X*ZX @ RETURN
3150 !
```

```
3160 ! **************  SUB TO CALC ERF,Q AND Z FROM KEYBOARD ********
3170 DISP "Enter X ([END LINE] to quit)";@ INPUT X$
3180   IF X$="" THEN 1570 ELSE X=VAL (X$) @ GOSUB 1180 @ GOSUB 1100
3190   PR$=" X= "&X$&"     ERF="&VAL$ (ERF)&"     Q= "
3200 PR$=PR$&VAL$ (Q)&"     Z= "&VAL$ (ZX) @ GOSUB 5920 @ GOTO 3170
3210 !
3220 ! ***********  SUB FOR SELECTION OF TYPE OF S/N ENTRY   ********
3230 PF=0 @ DISP "Select type of S/N entry: " @ DISP
3240 DISP "1 numeric"
3250 DISP "2 dB"
3260 DISP "3 R/Ro (S=N at Ro)"
3270 DISP "4 Range"
3280 DISP @ DISP "Enter #";@ INPUT SNFLG
3290 IF SNFLG<1 OR SNFLG>4 THEN DISP "Use # 1-4 " @ BEEP @ GOTO 3280
3300 IF SNFLG#4 THEN RETURN
3310 DISP "Enter range units, i.e. nmi, km, etc";
3320 INPUT RU$@ X$(4)="Range ("&RU$&")" @ RETURN
3330 !
3340 ! ***********  SUB TO ENTER NINT AND COLLAPSING RATIO(RHO) **********
3350 DISP "Enter no. of pulses integrated non-coherently, NINT (>=1) ";
3360 INPUT NINT@ IF NINT<1 THEN DISP "Use NINT>=1" @ DISP @ BEEP @ GOTO 3350
3370 IF FP (NINT) THEN DISP "Use NINT an integer" @ BEEP @ GOTO 3350
3380 DISP "Enter collapsing ratio, RHO,(>=1) ";@ INPUT R10@ R12=NINT*R10
3390 IF R10<1 THEN DISP "Use RHO>=1" @ DISP @ BEEP @ GOTO 3380
3400 S12=SQR (R12) @ RETURN
3410 !
3420 ! *************  SUB TO CALC IGF FROM KEYBOARD **************
3430 PR$="" @ GOSUB 5920
3440 DISP "Enter N (integer>=1)(0 to quit)";@ INPUT R12
3450   IF R12=0 THEN 1570
3460   IF FP (R12) THEN DISP "Use N an integer" @ BEEP @ GOTO 3440
3470   DISP "Enter Yb (>0)";@ INPUT R11
3480   IF R11<= 0 THEN DISP "Use Yb>0" @ DISP @ BEEP @ GOTO 3470
3490   Y=R11 @ N=R12 @ GOSUB 1250
3500   PR$="  N= "&VAL$ (N)&"     Yb= "&VAL$ (Y)&"    I="&VAL$ (IGF)
3510 GOSUB 5920 @ GOTO 3440
3520 !
3530 ! *************  SUB FOR KYBD CALC OF MARCUM'S Q FN **********
3540 DISP "Enter a,b (both >0) (0,0 to quit)";@ INPUT R4,R11
3550   IF R4=0 THEN 1570
3560   IF R4<0 OR R11<= 0 THEN DISP "Use a,b >0" @ DISP @ BEEP @ GOTO 3540
3570   TEM=R11 @ GOSUB 3660
3580   PR$="  a= "&VAL$ (R4)&"     b= "&VAL$ (R11)&"    Q(a,b)= "&VAL$ (Q)
3590 GOSUB 5920 @ GOTO 3540
3600 !
3610 ! ********  SUB FOR PARL ALGORITHM FOR MARCUM'S Q FUNCTION  *****
3620 ! Translation of PGM 3-2
3630 ! Entry for SUB 271, X=S/N
3640 R4=SQR (2*X) @ A=.5 @ X=1/H7 @ GOSUB 1020 ! a
3650 TEM=SQR (-(2*LOG (EE))) ! b
3660 RRZ=2/R4/TEM @ R2=R4/TEM
3670 R5,F1=1 @ IF R2>= 1 THEN R5,F1=0 @ R2=1/R2
```

```
3680 RT=R2 @ R1,R6,R8=0 @ R7=.5
3690 R1=R1+RRZ @ XX=R2+R1*R5+R6
3700   R6=R5 @ R5=XX @ R2=R2*RT
3710   XX=1+R1*R7+R8 @ R8=R7 @ R7=XX
3720 IF R7<1000 THEN 3690 ! can use larger number to increase accy
3730 X=-((R4-TEM)^2/2) @ GOSUB 720 @ Q=R5*EX/R7/2
3740 IF F1 THEN RETURN ELSE Q=1-Q @ RETURN
3750 !
3760 ! **************** SUB FOR PD CALC FOR CHI-SQ CASE    *********
3770 IF NFM#2 OR NL>1 THEN DISP "Chi-sq for S-S,1 look only!" @ RETURN
3780 CC=-(R12*R3/KCS) @ C=1/(1-CC)
3790 DEL1,SUMM=0 @ AJ,BJ,J=1
3800 IF R3>10*KCS THEN 3880 ! Branch to Eq. 15, Asymp. Series
3810 !   Eq. 14 Convergent series if Xbar/KCS <=10
3820 Y=R11 @ N=R12 @ GOSUB 1250 @ I=IGF @ DJ=R4 ! I OF Yb,N
3840 AJOLD=AJ*(1-I) @ SUMM=AJOLD+SUMM @ AJ=AJ*(1-C)*(KCS+J-1)/J
3850   I=I-DJ @ J=J+1 @ DJ=DJ*R11/(R12+J-1)
3860 IF ABS (AJOLD)>SUMM/1000 THEN 3840 ELSE PD=C^KCS*SUMM @ RETURN
3870 !
3880 ! ********* EQ. 15 ASYMPTOTIC SERIES IF Xbar/KCS>10 **********
3890 Y,YY=C*R11 @ N=KCS+1 @ GOSUB 1250 @ HJ=IGF @ EJ=R4*N/YY
3900 HJ=HJ+EJ @ DEL=BJ*(1-HJ) @ SUMM=SUMM+DEL
3910   BJ=(R12-N+J)*BJ/CC/J @ EJ=EJ*(N-J)/YY
3920   J=J+1 @ DEL2=DEL1 @ DEL1=DEL @ IF J<3 THEN 3900
3930   IF ABS (DEL1)>ABS (DEL2) THEN 3950 ! Asymp test
3940 IF ABS (DEL)>SUMM/1000 THEN 3900 ! Conv test
3950 PD=(SUMM-DEL1-DEL2)*(1-C)^(KCS-R12) @ RETURN
3960 !
3970 ! ************ SUB TO CALC PFR/Pfa FROM KEYBOARD *************
3980 DISP @ DISP "Enter M,N, for M-out-of-N detection (M<=N) (0,0 to quit)";
3990 INPUT M,NL@ IF M=0 THEN 1570
4000 IF M>NL THEN DISP "Use M<=N" @ BEEP @ GOTO 3980
4010 IF M<1 OR FP (M) THEN DISP "Use M an integer >0" @ BEEP @ GOTO 3980
4020 GOSUB 4220
4030 PR$="# of cells in correlation window="&VAL$ (KCOR) @ GOSUB 5920
4040 DISP @ DISP "Select operation" @ DISP
4050 DISP "1  PFR given Pfa"
4060 DISP "2  Pfa given PFR"
4070 DISP "Enter #";@ INPUT R@ IF R=2 THEN 4160
4080 IF R<1 OR R>2 THEN DISP "Use 1 or 2" @ BEEP @ GOTO 4070
4100 DISP "Enter Pfa (0 to quit)";@ INPUT PFA@ IF PFA=0 THEN 3980
4110   IF PFA<0 THEN DISP "Use Pfa>0 " @ DISP @ BEEP @ GOTO 4100 ELSE GOSUB 980
4120   PR$="  M="&VAL$ (M)&"   N= "&VAL$ (NL)&"  Pfa="&VAL$ (PFA)&"  PFR= "
4130 PR$=PR$&VAL$ (PFR) @ GOSUB 5920 @ GOTO 4100
4140 !
4150 ! **************** Pfa FROM PFR  FROM KEYBOARD******************
4160 DISP "Enter PFR (0 to quit)";@ INPUT PFR@ IF PFR=0 THEN 3980
4170   IF PFR<0 OR PFR>= 1 THEN DISP "Use 0<PFR<1" @ BEEP @ GOTO 4160
4180   GOSUB 4470 @ PR$="  M="&VAL$ (M)&"   N= "&VAL$ (NL)&"  PFR="
4190 PR$=PR$&VAL$ (PFR)&"  Pfa= "&VAL$ (PFA) @ GOSUB 5920 @ GOTO 4160
4200 !
```

```
4210 ! ************* CORRELATION WINDOW ENTRY ***********************
4220 DISP "Enter # of cells in correl window, KCOR (>=1)";@ INPUT KCOR
4230 IF KCOR<1 THEN DISP "Use KCOR>=1" @ DISP @ BEEP @ GOTO 4220 ELSE RETURN
4240 !
4250 ! ******* SUB TO ENTER DETECTION ALGORITHM AND FIND Pfa *******
4260 MNFLG=0 @ DISP "Enter # of looks per dwell, NL (>=1)";@ INPUT NL
4270 IF NL<1 THEN DISP "Use NL>=1 " @ DISP @ BEEP @ GOTO 4260
4280 IF FP (NL) THEN DISP "Use Integer only" @ BEEP @ GOTO 4260
4290 IF NL>1 THEN 4310
4300 M=1 @ DISP "Enter Pfa (0<Pfa<1) (Pfa=N*ln(2)/n)";@ INPUT PFA@ GOTO 4430
4305 !
4310 DISP "Enter # of detections per dwell for tgt rpt.,M";@ INPUT M
4320 IF M>NL THEN DISP "Use M<=NL" @ DISP @ BEEP @ GOTO 4310
4330 IF M<1 OR FP (M) THEN DISP "Try Again!" @ BEEP @ GOTO 4310
4340 GOSUB 4220 ! Cells in correlation window
4350 DISP "Enter Pfa, if known, or [EL] to calc Pfa from PFR";@ INPUT PFA$
4360 IF PFA$="" THEN 4400 ELSE PFA=VAL (PFA$)
4370 IF PFA<0 THEN DISP "Use Pfa>=0" @ DISP @ BEEP @ GOTO 4350
4380 GOSUB 980 @ GOTO 4430
4390 !
4400 DISP " Enter False Report Probability ";@ INPUT PFR
4410 IF PFR>= 1 OR PFR<= 0 THEN DISP "Use 0<PFR<1" @ DISP @ BEEP @ GOTO 4400
4420 GOSUB 4470 ! Calc. Pfa
4430 N=R12 @ nM,H7=N*LOG (2)/PFA ! Marcum's n
4440 GOSUB 640 @ R11=YB @ D11=10*LGT (YB)
4450 IF NL=1 THEN 5380 ELSE 5420 ! Print: N,n,RHO,Pfa,PFR,Yb
4460 !
4470 ! ********** SUB TO CALC Pfa GIVEN PFR, KCOR, M/N **************
4480 R4=PFR*KCOR ! Uses Newton-Raphson Root-Finding
4490 PFA=R4^(1/M) @ PR=PFR
4500 R5=PFA ! Pfa
4510  PFA=R5+.00000001 @ GOSUB 980 @ R3=PFR ! C(Pfa+1E-8)
4520  PFA=R5 @ GOSUB 980 @ R3=R3-PFR @ X=PR-PFR ! C(Pfa)
4530  PFA=X/100000000/R3+R5
4540 IF ABS (PFA-R5)>PFA/100 THEN 4500 ELSE PFR=PR @ RETURN
4550 !
4560 ! ********** SUB TO ENTER TARGET FLUCTUATION SPEED ***********
4570 C$(1)="Non-" @ C$(3)="Look-to-Look"
4580 C$(2)="Scan-to-Scan" @ C$(4)="Pulse-to-Pulse"
4590 D$=" Fluctuating Target"
4600 DISP @ FL=0 @ DISP "Select Fluctuation speed" @ DISP
4610 DISP "1 ";C$(1);D$
4620 DISP "2 ";C$(2);D$
4630 IF NL#1 THEN DISP "3 ";C$(3);D$
4640 DISP "4 ";C$(4);D$ @ DISP
4650 DISP "Enter Speed #";@ INPUT R
4660 IF R<1 OR R>4 THEN DISP "Use # 1-4" @ BEEP @ GOTO 4650
4670 IF NL=1 AND R=3 THEN DISP "Don't use #3 for 1 look" @ BEEP @ GOTO 4650
4680 PR$=C$(R)&D$ @ NFM=R @ GOTO 5950
4690 !
```

```
4700 ! ************* SUB TO ENTER TARGET FLUCT MODEL   ******************
4710 DISP "Select fluctuation Model" @ DISP
4720 IF NFM=4 THEN 4860 ! Branch for P-P Fluct,cont for S-S and L-L
4730 B$(1)="One-dominant, Swerling 3"
4740 B$(2)="Rayleigh, Swerling 1 "
4750 B$(3)="Log-normal"
4760 B$(4)="Generalized Exponential"
4770 B$(5)="Chi-square" ! Chi-sq, single look only
4780 ND=5-SGN (NL-1) @ FOR JJ=1 TO ND @ DISP JJ;" ";B$(JJ) @ NEXT JJ
4790 DISP "Enter Model #";@ INPUT FL
4800 IF FL<1 OR FL>ND THEN DISP "Use # 1-";ND @ DISP @ BEEP @ GOTO 4790
4810 PR$=B$(FL)&" Target " @ GOSUB 5950
4820 IF FL=3 OR FL=4 OR NL>1 THEN GOSUB 5470 ! Load non-fluct Pd curve
4830 ON FL GOTO 4840,4840,5740,5800,5860
4840 RETURN
4850 !
4860 ! ********* SUB TO ENTER TGT FLUCT MODEL FOR P-P FLUCT ************
4870 B$(1)="One-dominant, Swerling 4"
4880 B$(2)="Rayleigh, Swerling 2"
4890 DISP B$(2)
4900 DISP B$(1) @ FL=2
4910 DISP "Enter Swerling Model #";@ INPUT R@ IF R=2 THEN 4810
4920 FL=1 @ IF R=4 THEN 4810
4925 DISP "Use 2 or 4" @ BEEP @ GOTO 4910
4930 !
4940 ! ***************** SUB FOR G-L INTEGRATION (045) ****************
4950 YI=(X+1)*H5
4960 IF FL=4 THEN X,XX=YI^KGE @ GOSUB 720 @ X=XX/EX/YI*KGE @ GOTO 4990
4970 TEM=YI @ IF FL=3 THEN GOSUB 5040 @ GOSUB 1150 @ X=Z/SQRHO/TEM @ GOTO 4990
4980 X=-YI @ GOSUB 720 @ X=EX @ IF FL#2 THEN X=X*X*4*TEM
4990 TEM=X ! (80)
5000 R2=PD(JJ) @ JJ=JJ+1
5010 IF NFM#3 THEN PP=R2 @ MM=M @ N=NL @ GOSUB 770 @ R2=CP
5020 SUMM=SUMM+R2*TEM*WI(R4) @ RETURN ! R2=C(P) IF S-S OR P IF L-L
5030 !
5040 ! ******************   SUB FOR L-N MODEL (339) ***************
5050 X=LOG (YI*R13)/SQRHO @ RETURN
5060 !
5070 ! ********* SUB TO CALC PD FOR M-OUT-OF-N DETECTION   ********
5080 ! Also used for single look Gen Ex and Log-Normal
5090 ON NFM GOTO 5240,5110,5110,5250
5100 !
5110 H6=X99/R3 ! R3=S/N,H6=YMAX, S-S OR L-L Fluct.
5120 H5=H6/2 @ SUMM=0 @ JJ=1
5130 FOR R4=1 TO NI ! Gauss-Legendre integration
5140   X=ZI(R4) @ GOSUB 4940 @ X=-(ZI(R4)) @ GOSUB 4940
5150 NEXT R4
5160 IF FL=4 THEN X=-(H6^KGE) @ GOSUB 720 @ X=EX @ GOTO 5190 ! Gen'l exp. model
5170 IF FL=3 THEN YI=H6 @ GOSUB 5040 @ GOSUB 1100 @ X=Q @ GOTO 5190 ! L-N
5180 X=-H6 @ GOSUB 720 @ X=EX @ IF FL#2 THEN X=X*X*(2*H6+1)
5190 PD=X+SUMM*H5 ! P
5200 IF NFM#2 THEN PP=PD @ MM=M @ N=NL @ GOSUB 770 @ PD=CP
5210 GOTO 2620 ! PRINT Pd
5220 !
```

```
5230 ! ********  select fluct tgt sub for Marcum and P-P fluct tgt ****
5240 GOSUB 3060 @ GOTO 5200 ! Marcum case
5250 ON FL GOSUB 2910,2790 @ GOTO 5200 ! Swerling 4 or 2
5260 !
5270 ! *** SUB FROM PGM 3-4 TO CALC Do(N) FOR Pd=0.99, NON-FLUCT TGT ***
5280 X=PFA @ GOSUB 1210 @ TEM=IQ @ TT=T @ X=.99 @ GOSUB 1210 ! INV Q(0.99)
5290 IF R12=1 THEN DZN=((TT-IQ)^2-1)/2 @ RETURN ELSE DC1=(TEM-IQ)^2/2
5300 DZN=(1+SQR (1+9.2*R12/DC1))*DC1/2/NINT @ RETURN
5310 !
5320 ! ***************  KEYBOARD CALC OF INVERSE Q ***************
5330 DISP "Enter Q(X) (0<Q<1) (0 to quit)";@ INPUT X@ IF X=0 THEN 1570
5340 IF X<0 OR X>= 1 THEN DISP "Use 0<Q(X)<1" @ DISP @ BEEP @ GOTO 5330
5350 GOSUB 1210 @ PR$="Q("&VAL$ (IQ)&")="&VAL$ (X) @ GOSUB 5920 @ GOTO 5330
5360 !
5370 ! ***************  PRINT/DISP, Single look  Pfa, Yb,N ***************
5380 PR$="Single Look, Pfa="&VAL$ (PFA)&", N="&VAL$ (NINT)&", RHO="&VAL$ (R10)
5390 PR$=PR$&", Yb="&VAL$ (FNRD(R11))&", Yb(dB)="&VAL$ (FNRD(D11)) @ GOTO 5920
5400 !
5410 ! *************  PRINT/DISP, Multiple look, Pfa, Yb, N, KCOR ********
5420 PR$=VAL$ (M)&" Detects-out-of "&VAL$ (NL)&" looks: Pfa="&VAL$ (FNRE(PFA))
5430 PR$=PR$&", PFR="&VAL$ (FNRE(PFR))&", N="&VAL$ (NINT)&", RHO="&VAL$ (R10)
5440 GOSUB 5920 @ PR$=" KCOR="&VAL$ (KCOR) @ GOTO 5390
5450 !
5460 ! ***********************  SINGLE LOOK CURVE  *****************
5470 DISP "Select Non-fluctuating single-look curve"
5480 DISP "1 Marcum"
5490 DISP "2 User-defined"
5500 IF MNFLG=1 THEN DISP "3 Same as before"
5510 DISP @ DISP "Enter # ";@ INPUT R
5520 IF MNFLG=1 AND R=3 THEN RETURN ! RETURN, use old curve
5530 IF R#2 THEN 5550 ELSE PR$="User-defined" @ DISP "Enter name for curve, ";
5540 DISP "up to 1 line";@ INPUT UD$@ DISP "Enter X99";@ INPUT X99@ GOTO 5590
5550 IF R#1 THEN DISP "Use # 1-";2+MNFLG @ DISP @ BEEP @ GOTO 5510
5560 GOSUB 5270 @ X99=DZN/R10 @ PR$="Marcum"
5570 ! NOTE X99 is divided by RHO to be compatible with the input S/N which
5580 !  is also divided by RHO for the canned curves for Sw 1-5
5590 IF PN#1 THEN PRINT PR$;" single look, non-fluctuating detection curve"
5600 IF PN#1 AND R=2 THEN PRINT UD$
5610 MNFLG,JJ=1 @ PR$="X99= "&VAL$ (DZN) @ GOSUB 5950
5620 FOR J=1 TO NI
5630   T=1 @ GOSUB 5670 @ T=-1 @ GOSUB 5670
5640 NEXT J @ PR$="" @ GOTO 5920
5650 !
5660 ! ************  SUB TO LOAD NON-FLUCT CURVE  *****************
5670 R3=X99*(1+T*ZI(J))/2 @ RP=R3*R10
5680 IF R=1 THEN GOSUB 3060 @ GOTO 5710
5690 DISP "P(";RP;")= ";@ INPUT PD
5700 IF PD<0 OR PD>1 THEN DISP "Use 0<Pd<1" @ BEEP @ GOTO 5690
5710 PD(JJ)=PD @ JJ=JJ+1 @ PR$="  X= "&VAL$ (RP)&"  PD= "&VAL$ (PD) @ GOTO 5920
5720 !
5730 ! ************  L-N PARAMETER ENTRY  ********************
5740 DISP "Enter L-N Parameter, RHO (>1)";@ INPUT R13
5750 IF R13<= 1 THEN DISP "Use RHO>1" @ DISP @ BEEP @ GOTO 5740
5760 SQRHO=SQR (2*LOG (R13)) @ IF PN#1 THEN PRINT "RHO=";R13
5770 RETURN
5780 !
```

```
5790 ! ******** GEN EXP PARAMETER ENTRY *********************
5800 DISP "Enter Generalized Exp. Par., KGE";@ INPUT KGE
5810 IF KGE<= 0 THEN DISP "Use KGE>0" @ DISP @ BEEP @ GOTO 5800
5820 IF PN#1 THEN PRINT "KGE=";KGE
5830 RETURN
5840 !
5850 ! **************** DUO-DEGREES OF FREEDOM ENTRY ************
5860 DISP "Enter Duo-degrees of Freedom, KCS";@ INPUT KCS
5870 IF KCS<= 0 THEN DISP "Use KCS>0" @ DISP @ BEEP @ GOTO 5860
5880 IF PN#1 THEN PRINT "KCS=";KCS
5890 RETURN
5900 !
5910 ! **************** DISPLAY/PRINT SUBS *********************
5920 DISP PR$ @ IF PN#1 THEN PRINT PR$
5930 RETURN
5940 !
5950 DISP PR$ @ DISP @ IF PN#1 THEN PRINT PR$ @ PRINT
5960 RETURN
5970 !
5980 DISP USING F1 ; SS,PD @ IF PN#1 THEN PRINT USING F1 ; SS,PD
5990 RETURN
6000 !
6010 DISP USING F2 ; SS,PD @ IF PN#1 THEN PRINT USING F2 ; SS,PD
6020 RETURN
6030 !
6040 DISP USING F3 ; SS,PD @ IF PN#1 THEN PRINT USING F3 ; SS,PD
6050 RETURN
6090 !
6100 CHAIN "Autost" ! return to disc menu
6110 !
6120 ! ************ FACTORIAL N  from keyboard *********************
6130 DISP @ DISP "CAUTION!! Factorials calculated by this program do not have "
6140 DISP "  full 12 place accuracy and are for positive values only. "
6150 DISP "  The binary program MATHBIg provides full accuracy for both"
6160 DISP "  positive and negative numbers."
6170 DISP @ DISP "Enter N (0 to quit)";@ INPUT N@ IF N=0 THEN 1590
6180   IF N>10000000000 THEN DISP "N too large" @ BEEP @ GOTO 6170
6190   IF N<0 THEN DISP "Use N>0" @ BEEP @ GOTO 6170
6200   GOSUB 890 @ X=FACT @ PR$=VAL$ (N)&" ! = "
6210   IF OFFLAG=0 THEN PR$=PR$&VAL$ (X) @ GOTO 6240
6220   X=X/LOG (10) ! LOG base 10 for factorial in log format
6230   PR$=PR$&VAL$ (10^FP (X))&" E "&VAL$ (INT (X))
6240 GOSUB 5920 @ GOTO 6170
6250 !
6260 ! *********** INITIALIZATION &  PRINTER SELECTION ***************
6270 PAGESIZE 24 @ CLEAR @ DEG @ PRINTER IS 1 @ IOFLAG=0
6280 DISP TAB (10);T$ @ DISP
6290 DISP "Enter today's date (any comma-less format)";@ INPUT DA$
6300 DISP "Select Printer:" @ DISP
```

```
6310 DISP "1   CRT"
6320 DISP "2   HP-82905B--HP-IB"
6330 DISP "3   HP-82905B--Centronics"
6340 DISP "4   MX-100"
6350 DISP "Enter #";@ INPUT PN
6360 IF PN<1 OR PN>4 THEN DISP "Use # 1-4 " @ BEEP @ GOTO 6350
6365 IF PN#1 THEN DISP "Make sure Top-Of-Form is set properly, Press [END
LINE]";@ INPUT PR$
6370 ON PN GOTO 6470,6390,6400,6440
6380 !
6390 PRINTER IS 301 @ ZZZ=3 @ GOTO 6410
6400 PRINTER IS 701 @ ZZZ=7
6410 ON ERROR GOTO 6430 @ a=BTD ("1") ! I/O ROM TEST
6420 CONVERT OUT ZZZ PAIRS ; ZZZ$@ IOFLAG=1
6430 OFF ERROR @ PRINT CHR$ (27)&"&l1L" @ GOTO 6460
6440 PRINTER IS 701 @ PRINT CHR$ (27);"N";CHR$ (8)
6460 PRINT TAB (15);T$ @ PRINT TAB (25);DA$
6470 DISP @ DISP "Answer questions with 'Y' or 'N', [END LINE]" @ DISP
6480 RETURN
6490 !
6500 ! ****************** PLOT DATA  *******************
6510 ! INPUTS : XL=MIN X, XR=MAX X, YT=MAX Y, YB=MIN Y
6550 DISP "Enter LINE TYPE no. 1-8 " @ INPUT LT
6560   IF PA=1 THEN LIMIT 0,125,0,75 @ X2=155 @ X3=85 @ GRAPH @ GOTO 6590
6570 IF IOFLAG THEN CONVERT OUT ZZZ ! Stop 0 conv for ext plotter
6580 LIMIT 12,252,18,192 @ X2=134 @ X3=73
6590 IF AP=0 THEN GOSUB 6860
6600 LOCATE 18,X2,10,84 @ SETGU
6610 IF AP=1 THEN 6740
6620 MOVE 18,89 @ CSIZE 4 @ LDIR 0 @ LORG 1
6630 IF NL=1 THEN LABEL USING P4 ; R10 ELSE LABEL USING P2 ; M,NL,KCOR,R10
6640 P4: IMAGE "Single look detection, RHO=",2D
6650 P2: IMAGE D,"-out-of-",D," detection,  ",DD," cell window, RHO=",2D
6660 MOVE 18,87 @ LORG 3
6670 IF NL=1 THEN LABEL USING P5 ; PFA,H7,NINT @ GOTO 6690
6680 LABEL USING P1 ; PFA,PFR,H7,NINT
6690 P5: IMAGE "Pfa=",D.DDE,"  n=",D.3DE,"  N=",5D
6700 P1: IMAGE "Pfa=",D.DDE,"  PFR=",D.DDE,"  n=",D.3DE,"  N=",5D
6710 MOVE 18,5 @ LABEL USING P3 ; R11
6720 P3: IMAGE "Yb=",3D.DD
6730 MOVE X2,5 @ LORG 9 @ LABEL DA$
6740 SCALE XL,XR,YB,YT
6750 MOVE XL+DXS,FF(1) @ I=1 @ LINE TYPE LT
6760 FOR X=XL+DXS+DX TO XR STEP DX
6770   I=I+1 @ DRAW X,FF(I)
6780 NEXT X @ IF PA=1 THEN 6800 ELSE PEN 0 @ MOVE XR,YT
6790 IF IOFLAG THEN CONVERT OUT ZZZ PAIRS ; ZZZ$ ! Reactivate 0 conv.
6795 RETURN
6800 ALPHA @ DISP "Dump to Printer";@ INPUT R$
6810 R$=UPC$ (R$) @ IF R$="N" THEN RETURN
6820 IF R$#"Y" THEN DISP "Y or N?" @ BEEP @ GOTO 6800
```

```
6830 IF PN=1 THEN GOSUB 6300 @ GOTO 6830 ! Select printer (not CRT)
6840 IF PN=4 THEN DUMP GRAPHICS 0,0,0,-1 @ RETURN ELSE DUMP GRAPHICS @ RETURN
6850 !
6860 IF PA#1 THEN LIMIT 12,252,18,192 @ X2=134 @ X3=73 @ GOTO 6880
6870 LIMIT 0,125,0,75 @ X2=155 @ X3=85 @ GRAPH @ GCLEAR @ FRAME
6880 LOCATE 18,X2,10,84 @ FRAME
6890 MOVE 4,50 @ LORG 4 @ LDIR 90 @ CSIZE 5,.7,30 @ LABEL YL$
6900 MOVE X3,94 @ CSIZE 8 @ LDIR 0 @ LABEL T$
6910 MOVE X3,5 @ LORG 6 @ CSIZE 5 @ LABEL X$(SNFLG)
6920 NDX=0 @ SCALE XL,XR,YB,YT @ CSIZE 4
6930 ! ******  select minor TIC spacing and decimal in labels   *******
6940 X=ABS (XL-XR) @ IF X<= 2.5 THEN DELX=.05 @ NDX=2 @ GOTO 7020
6950 IF X<= 5 THEN DELX=.1 @ NDX=1 @ GOTO 7020
6960 IF X<= 10 THEN DELX=.2 @ GOTO 7020
6970 IF X<= 25 THEN DELX=.5 @ NDX=1 @ GOTO 7020
6980 IF X<= 50 THEN DELX=1 @ GOTO 7020
6990 IF X<= 100 THEN DELX=2 @ GOTO 7020
7000 IF X<= 250 THEN DELX=4 @ GOTO 7020
7010 IF X<= 500 THEN DELX=10 ELSE DELX=20
7020 FXD NDX,1 @ LGRID -DELX,.02,XL,0,5,5 @ RETURN
7030 !
7040 ! ********** Sub to Select plotter, load binary GDUMP ******
7050 DISP "Enter Plotter Address (CRT=1)";@ INPUT PA
7060 IF PA<1 THEN DISP "Use # >=1" @ BEEP @ GOTO 7050
7070 ON ERROR GOTO 7090 @ GDUMP @ GOTO 7140 ! Error if GDUMP not resident,
7080 !   displays "Copyright Hewlett-Packard 203.09", if resident
7090 ON ERROR GOTO 7120 @ LOADBIN "GDUMP" @ GOTO 7070 ! several errors possible
7100 ! ERRN=67 indicates GDUMP not on disc
7110 ! ERRN=130    "      bad disc, no disc, door open, etc.
7120 IF ERRN =67 OR ERRN =130 THEN DISP "Insert disc with binary GDUMP, press
[CONT]" @ PAUSE @ GOTO 7090
7130 DISP "Unknown error,ERRN=";ERRN ;" try again" @ GOTO 7050
7140 OFF ERROR @ PLOTTER IS PA @ RETURN
```

Chapter 4
Signal-to-Noise Calculations
Program SIG/CLUT

```
10  ! Pgm SIG/CLUT 2/28/84  SURFACE AND VOLUME CLUTTER, and S/N,S/J, J/N
20  ! Written by W. Skillman, combo of Pgms. 4-1,2,3 and 4 of "Radar Calc..
30  !
40  ! Improvements vs. Radar Calc...
50  !    Sums main beam surf clut from <=45 deg. dep angle to
60  !       radar horizon if PRF is ambiguous
70  !    Bistatic beam crossover clutter reduction added to surface &
80  !       volume clutter (no effect on surface for vertical ant. separation)
90  !    Separate Tx Beamwidth used for non-scanning case only
100 !       for surface clutter
110 !    Selectable Elevation sidelobe level for surface clutter
120 !    Integrated volume and surface clutter at IF for CW only
130 !    Doppler proc gain
140 !    Pd vs S/N curve for Pd vs range output
150 !    Selectable range and altitude units
160 !    Separate clutter, target and jammer losses
170 !
180 ! Important variables: (All angles in degrees)
190 !    A$  = array of radar parameter titles(indexed like P)
200 !    AA  = 1+hR/Re
210 !    AL = Grazing angle array
220 !    ALPHM = max grazing angle for U-D clut model
230 !    BB  = hR(2Re+hR)
240 !    C=J/N Constant (see Pgm 4-4)
250 !    C2 = velocity of light/2 (nmi/sec)
260 !    CFLAG: 1=const sigma-zero,  2=const gamma, 3=User-defined
270 !    CM = Clutter coefficient (dB)
280 !    CTON = single patch clutter (dB)
290 !    CTOUT= C/N (dB) in output bandwidth
300 !    DBB = Dep of Bottom of Beam
310 !    DD = Dep of bottom of rain
320 !    DELD = Effective Elev. Beamwidth, rain clutter
330 !    DELR = IPP (range units)
340 !    DEP = Depression angle array
350 !    DHOR = Dep of Radar Horiz.
360 !    DTB = Dep. of Top of Beam
370 !    DTRN = Dep of Top of Rain
380 !    DUTY = duty cycle
390 !    FAC = patch geom. factor array
400 !    FF = output C/N array (dB)
410 !    FFLAG: 0=IPP >= horizon range, 1=otherwise
420 !    F1,F4,F5, Rain clutter flags(see Pgm 4-3)
430 !    GRJDB=Radar rx gain in jammer direction (dB)
440 !    GSL = elev sidelobe to main beam ratio (power)
450 !    HI = Index for alt units
460 !    HRAD = radar height (range units)
470 !    HRAIN = Rain height    "      "
480 !    IOFLAG = 1 if HP-82905 Printer used and I/O ROM installed, else =0
490 !    JN = Jam-to-noise ratio
```

```
500 !   K1 = clutter constant
510 !   K1T = same as K1 but for target
520 !   NG = # gates in IPP
530 !   P  = radar parameter array
540 !       1=Xmtr peak power (watts)        15=Rx Elev beamwidth
550 !       2=Pulse comp. ratio              16=Tx Bm Start Elev
560 !       3=Transmit gain (dB)             17=Rx  "    "     "
570 !       4=Receive   "    "               18=Elev scan rate (deg/sec)
580 !       5=Wavelength(cm)                 19=PRF (Hz)
590 !       6=Noise temp. (deg K)            20=ambig. range (gates)
600 !       7=Clutter Losses (dB)            21=duty cycle
610 !       8=Xmt Bandwidth (Hz)             22=no. gates in IPP
620 !       9=Doppler Proc. gain (dB)        23=constant K1
630 !      10=Azimuth beamwidth              24=C/N (dB)
640 !      11=Sigma-zero (dB)                25=C/N output (dB)
650 !      12=Gamma (dB)                     26=radar horiz range(R units)
660 !      13=Radar altitude(alt units)     27=Rainfall rate (mm/hr)
670 !      14=Compressed pulse width(usec)   28=Backscatter coef (m2/m3)
680 !      30=Beam Elev                      29=Rain height (alt units)
690 !      31=TX Elev. Beamwidth             32=Elev Sidelobe gain(dB)
700 !      33=Jammer losses(dB)              34=ERPD (kw/MHz)
710 !      35=Jammer bandwidth (Hz)          36=Az sidelobe gain (dB)
720 !      37=standoff jammer rng(R units)   38=bistatic illum. range(R units)
730 !      39=Antenna separation (ft)        40= target loss(dB)
740 !
750 !   PA = plotter address (See sub 9000)
760 !   PD = Pd vs S/N (dB) array
770 !   PN = printer code (see sub 7350)
780 !   RAMB = ambig range of range gate (range units)
790 !   RC = beam crossover range            "
800 !   RE = Effective Earth's radius =4588nmi (4/3 actual of 3440nmi)
810 !        (Converted to RANGE units in sub 6950)
820 !   RES=Range with Escort Jammer (Range units)
830 !   RHOR = radar horizon range-4/3 Earth rad mod (range units)
840 !   RHORRN =rain horiz range (R units)
850 !   RI = Index for range units; 1=km,2=nmi,3=mi, 4=ft
860 !   RR = clutter patch range (R units)
870 !   RRS = start range for clutter patch sum       (Range units)
880 !   RZ =Range where S/N=1, Marcum's idealized range   "    "
890 !   RSO=Range with Standoff Jammer              "    "
900 !   RTB=Clear Range perf, Bistatic radar        "    "
910 !   RTM=Clear Range perf, Monostatic radar      "    "
920 !   RVMIN= min range for chaff clutter          "    "
930 !   SEP$ = "V" for vertically separated antennas, "H" for horiz.
940 !   SFLAG: 2=surface clut, 3= surf clut +min RCS,4=surf clut +S/N,S/N+C
950 !   SIGT= target RCS (sq. meter )
960 !   SN = signal-to-noise (power)
970 !   SUMC = clutter sum (power)
980 !   THR = angle from receive beam peak
990 !   THT =  same for transmit  beam
1000 !    TNDB = S/N for detection (dB)
```

```
1010 !    UD = User-defined clutter model array (dB)
1020 !    UU = Conversion factor from nmi to user-defined units
1030 !    UU$ =abbrev. for user-defined range and alt. units
1040 !    ZZZ = HP-82905 Printer interface code
1050 !    ZZZ$ = Normal zero to slashed zero conversion string
1060 !
1070 DATA Xmtr. peak power (w),Pulse compression ratio,Transmit gain (dB)
1080 DATA Receive gain (dB),Wavelength (cm),Noise Temp. (deg K)
1090 DATA Losses (dB),Xmt.(or IF) Bandwidth (Hz),Doppler Processing gain (dB)
1100 DATA Azimuth Beamwidth (deg),Sigma Zero (dB),Gamma (dB)
1110 DATA Radar Altitude,Compressed Pulsewidth (us),Rx Elev. BW (deg)
1120 DATA Transmit Beam Start El angle(+ UP)(deg),Receive Beam start angle (deg)
1130 DATA Elevation scan rate (deg/sec) (+up),PRF (Hz),Ambiguous range in gates
(0=trans),duty cycle,number of contiguous gates,constant K1
1140 DATA I.F.clutter to noise (dB),Output C/N (dB),Radar Horizon Range
1150 DATA Rainfall rate (mm/hr),Backscatter coefficient (sq.m./cu.m.)
1160 DATA Rain height,Beam Elevation (deg),Tx elev.Beamwidth (deg)
1170 DATA Elev. Sidelobe Gain (dB) WRT Peak,Jammer losses (dB)
1180 DATA Jammer ERPD (kw/MHz)
1190 DATA Jammer bandwidth (Hz),Az. Sidelobe Gain (dB) WRT Peak
1200 DATA Standoff Jammer range,Bistatic Illum Range,Antenna Separation(ft)
1210 DATA target loss (dB)
1220 DATA Constant Sigma-zero,Constant gamma,User-defined,Clutter model
1230 DATA (km),1.852,(nmi),1,(mi),1.15078,(ft),6076.115,80938
1240 !
1250 DIM A$(40)[40],P(40),T$[35],S$[40],X$[30],UD(20),UU$(4),UU(4),PP$[25]
1260 DIM DEP(100),FAC(100),FF(300),M$(4)[25],PR$[80],AL(100),PD(100)
1270 ZZZ$="0"&CHR$ (210) @ T$="** SURFACE/VOLUME CLUTTER **" @ GOTO 1470
1280 !
1290 ! *****************  DEFINED FUNCTIONS  ********
1300 ! Monostatic range , peak of beam,clear env.
1310 DEF FNRM(SIGT,TNDB) = (310.6*K1T*FNIPDB(P(9)-TNDB)*SIGT)^.25*UU(RI)
1320 ! Jammer constant
1330 DEF FNCC(ERPD) = 6692910*FNIPDB(P(4)-P(33))*ERPD*P(5)^2/P(6)
1340 DEF FNPDB(X) = 10*LGT (MAX (EPS ,ABS (X))) ! Power in dB
1350 DEF FNIPDB(X) = 10^(X/10) ! dB to power ratio
1360 DEF FNRD(X) = INT (1000*X+.5)/1000 ! Round to 3 dec.
1370 !
1380 ! FNGAIN(th,bw) ! Gaussian beamshape,th=ang from pk, bw is bmwdth
1390 X=th/bw @ IF ABS (X)>XMAX THEN GAIN=GSL ELSE GAIN=16^(-(X*X))
1400 RETURN ! power is WRT peak.
1410 !
1420 ! FNDEP(RR,H) ! Calc dep ang(+down),given tgt slant range RR& height,H
1430 DEL=HRAD-H ! HRAD is radar height, RE is 4/3rds Earth rad.
1440 IF R<ABS (DEL) THEN DEP=90*SGN (DEL) @ RETURN
1450 DEP=ASN ((RR/RE+(AA*AA-(1+H/RE)^2)*RE/RR)/2/AA) @ RETURN
1460 !
1470 P(11),P(12),RROLD,CFLAGOLD=0 @ DR$=CHR$ (7)
1480 GOSUB 7280 @ GOSUB 7520 ! Init, Printer & plotter selection, load GDUMP
1490 FOR I=1 TO 40 @ READ A$(I)@ NEXT I @ YL$=A$(25)
1500 FOR I=1 TO 4 @ READ M$(I)@ NEXT I
```

```
1510 FOR I=1 TO 4 @ READ UU$(I),UU(I)@ NEXT I @ CLEAR
1520 READ C2@ FOR I=1 TO 9 @ GOSUB 2870 @ NEXT I ! Enter Common parameters
1530 GOSUB 3800 ! Calc. constant, K1
1540 GOSUB 6870 ! Select range,ht units
1550 !
1560 DISP @ DISP "  ** MAIN MENU **   Select Option:" @ DISP
1570 DISP "1 New Parameters"
1580 DISP "2 Surface Clutter"
1590 DISP "3 Rain/Chaff  Clutter"
1600 DISP "4 S/N,J/N"
1610 DISP "5 New Range, height units"
1620 DISP "6 Disc Menu" @ DISP
1630 DISP "Enter #";@ INPUT R@ GFLAG=0 @ IF R=6 THEN CHAIN "Autost"
1640 IF R<1 OR R>6 THEN DISP "USE # 1-6" @ BEEP @ GOTO 1630
1650 ON R GOTO 1670,1700,4990,6140,1540
1660 !
1670 NN=9 @ GOSUB 3500 @ GOTO 1560
1680 !
1690 ! ***************  SURFACE CLUTTER  *************************
1700 T$="**  Surface Clutter  **" @ I=10 @ GOSUB 2870 ! Azimuth Beamwidth
1710 GOSUB 3710 ! select clutter model
1720 I=13 @ GOSUB 2860 ! altitude
1730 GOSUB 3960 ! calc horizon range(RHOR),AA,BB
1740 FOR I=14 TO 15 @ GOSUB 2870 @ NEXT I ! pulsewdth,rx.elev. bw
1750 I=31 @ GOSUB 2870 ! Trans. Elev. Beamwidth
1760 I=32 @ GOSUB 2870 @ GOSUB 6090 ! Elev sidelobe level
1770 I=39 @ GOSUB 2870 @ GOSUB 7180 ! Ant sep, beam xover range
1780 FOR I=16 TO 19 @ GOSUB 2870 @ NEXT I ! tx,rx st angles,rate,PRF
1790 RHORRN=0 @ GOSUB 4010 ! calc DUTY,DELR,NG, set FFLAG if ambig PRF
1800 !
1810 DISP T$;" Menu: Select next operation" @ DISP
1820 DISP "1 New Parameter(s)"
1830 DISP "2 Surface Clutter"
1840 IF FFLAG THEN 1870
1850 DISP "3 Surface Clutter + min RCS"
1860 DISP "4 Surface Clutter +S/N, S/(N+C)"
1870 DISP "5 MAIN MENU" @ DISP
1880 DISP "Enter #";@ INPUT R@ SFLAG=R
1890 IF R<1 OR R>5 THEN DISP "use # 1-5" @ BEEP @ GOTO 1880
1900 IF FFLAG AND (R=3 OR R=4) THEN DISP "Use # 1,2 or 5" @ BEEP @ GOTO 1880
1910 ON R GOTO 2000,2030,1940,1970,1560
1920 !
1930 ! **********  T/N and tgt loss for min  RCS  ***********
1940 GOSUB 4850 @ GOTO 2030
1950 !
1960 ! **********  RCS and tgt loss for S/N, S/(N+C) ********
1970 GOSUB 4900 @ GOTO 2030
1980 !
1990 ! ***********  change surf clutter parameters  *********
2000 NN=20 @ GOSUB 3500 @ GOTO 1810
2010 !
```

```
2020 ! *************    START SURFACE CLUTTER CALC   ***********
2030 J=1 @ IF FFLAG=0 THEN 4300 ! branch if unambig range PRF
2040 DISP "Ambiguous PRF, select sweep mode" @ DISP
2050 DISP "1 C/N vs Elev"
2060 DISP "2 C/N vs Range (in res cells or gates)" @ DISP
2070 DISP "Enter #" @ INPUT R@ IF R=2 THEN 2270
2080 IF R#1 THEN DISP "Use 1 or 2 " @ BEEP @ GOTO 2070
2090 DISP "  Assuming gate spacing=width, 0=Tx pulse, ";
2100 I=20 @ GOSUB 2870 ! Range gate #
2110 IF P(20)>NG THEN DISP "Use gate # <=";NG @ BEEP @ GOTO 2100
2120 DISP "Enter min, max and ";DR$;"elev.(deg) (+ up)";
2130 INPUT EL1,EL2,DEL@ NP=P(20) @ DE=EL2-EL1
2140 IF DE<0 THEN DISP "Use min<max" @ BEEP @ GOTO 2120
2150 IF DEL>DE OR DEL<= 0 THEN DISP "Use 0<";DR$;"elev<";DE @ BEEP @ GOTO 2120
2155 IF DE>300*DEL THEN DISP "Too many points for FF array" @ BEEP @ GOTO 2120
2160 PR$="" @ GOSUB 3920
2170 PR$="    Elev (deg)        C/N (dB) " @ GOSUB 3920
2180 FOR EL=EL1 TO EL2 STEP DEL @ P(16),P(17)=EL
2190    GOSUB 2630 @ FF(J)=CTOUT @ J=J+1 ! Sum over patches to RHOR
2200 NEXT EL
2210 DISP "PLOT";@ INPUT R$@ R$=UPC$ (R$) @ IF R$="N" THEN 1810
2220 IF R$#"Y" THEN DISP "Y or N?" @ BEEP @ GOTO 2210
2230 XL=EL1 @ XR=EL2 @ DX=DEL @ DXS=0 @ X$="Beam Elevation (deg)"
2240.TL$="Gate="&VAL$ (P(20)) @ TR$="" @ GOTO 4430 ! Plot C/N vs. Elev
2250 !
2260 ! *************   SURFACE C/N VS IPP  SWEEP   *****************
2270 DISP "Enter min & max range gate # and ";DR$;"gates";
2280 INPUT N1,N2,DNP@ SUMOUT=0
2290 IF N2>NG OR DNP>NG THEN DISP "use # <=";NG @ BEEP @ GOTO 2270
2300 IF N2<= N1 OR DNP<= 0 THEN DISP "Use min<max" @ BEEP @ GOTO 2270
2305 IF N2-N1>300*DNP THEN DISP "Too many pts for FF array" @ BEEP @ GOTO 2270
2310 PR$="" @ GOSUB 3920
2320 PR$="    Gate #               C/N(dB)" @ GOSUB 3920
2330 FOR NP=N1 TO N2 STEP DNP
2340    EL=NP @ GOSUB 2630 @ FF(J)=CTOUT @ J=J+1 ! Sum over patches to RHOR
2350    SUMOUT=SUMOUT+FNIPDB(CTOUT) ! Sum over IPP for CW I.F. C/N
2360    RRS=RRS+DNP*DELR
2370 NEXT NP
2380 ! Calc. C/N at I.F. for CW case only
2390 X,CNIFDB=FNPDB(SUMOUT/(J-1))-P(9)
2400 PR$="" @ GOSUB 3920
2410 PR$="C/N,IF,CW (dB)" @ GOSUB 3860
2420 DISP "PLOT";@ INPUT R$@ R$=UPC$ (R$) @ IF R$="N" THEN 1810
2430 IF R$#"Y" THEN DISP "Y or N?" @ BEEP @ GOTO 2420
2440 XL=0 @ XR=N2 @ DX=DNP @ DXS=N1 @ X$="Range (res. cells)"
2450 GOTO 4420 ! Plot C/N vs IPP
2460 !
2470 ! ********* INITIALIZE FOR  SURFACE CLUTTER SUM  ************
2480 RAMB=NP*P(14)*UU(RI)*C2/1000000 ! Ambiguous range
2490 D=HRAD*SQR (2) ! Starts at 45 deg dep
2500 NPS=0 @ IF D>RAMB THEN NPS=1+INT (D/DELR)
2510 RR,RRS=RAMB+NPS*DELR
2520 IF RR=RROLD THEN RETURN ELSE RROLD=RR @ I=1 ! Store Dep,Range Factor
2530 GOSUB 2560 @ RR=RR+DELR @ I=I+1 @ IF RR<RHOR THEN 2530 ELSE RETURN
```

```
2540 !
2550 ! ****************** calc dep, graz & FAC *************
2560 GOSUB 2720 ! calc Dep, Graz
2570 F=1 @ IF CFLAG=2 THEN F=X
2580 IF RC=0 OR SEP$="V" THEN 2600
2590 IF RR<RC THEN FAC(I)=0 @ RETURN
2600 F=F*(1-RC/RR) @ X=RR/UU(RI) @ FAC(I)=F/X/X/X/Y @ RETURN
2610 !
2620 ! ******* SUM CLUTTER FROM DEP =45 TO HORIZON  ***************
2630 GOSUB 2480 @ I,SUMC=0 @ RR=RRS @ NT=NPS
2640 I=I+1 @ GOSUB 2760 @ RR=RR+DELR @ SUMC=SUMC+X @ NT=NT+1
2650 IF RR<RHOR THEN 2640
2660 X=K1*P(10)*P(14)*SUMC
2670 CTOUT=61.77+FNPDB(X)+CM+P(9) ! DFT Output C/N (dB)
2680 DISP TAB (3);EL,CTOUT @ IF PN#1 THEN PRINT TAB (3);EL,CTOUT
2690 RETURN
2700 !
2710 ! ****** SUB TO CALC DEP AND GRAZ ang, & for surf clut *****
2720 X=(BB/RR-RR)/2/RE
2730 AL(I)=ASN (X) @ Y=SQR (1-X*X) @ DEP(I)=ACS (Y/AA) @ RETURN
2740 !
2750 ! *********** CALC CLUTTER FACTOR, ONE PATCH ************
2760 THR=P(17)+DEP(I) ! initial angle Rx beam to LOS to patch
2770 THT=P(16)+DEP(I) ! ditto, Tx
2780 IF P(18)=0 THEN 2810 ! Branch for non-scanning case
2790 THR=THR+P(18)*(NP/P(19)+RR/UU(RI)/C2) !  Scanning case
2800 THT=THT+P(18)*(NP-NT)/P(19)
2810 th=THR @ bw=P(15) @ GOSUB 1380 @ GR=GAIN
2820 th=THT @ bw=P(31) @ GOSUB 1380 @ X=GR*GAIN*FAC(I) @ RETURN
2830 !
2840 ! ****************** SUB TO INPUT DATA  *******************
2850 V$=UU$(RI) @ GOTO 2880 ! Add Range unit notation
2860 V$=UU$(HI) @ GOTO 2880 ! Ditto for Alt.
2870 V$=""
2880 DISP "Enter ";A$(I);V$;@ INPUT X@ P(I)=X @ PR$=A$(I) @ GOTO 3880
2890 !
2900 ! *********** User-defined clutter model interpolation ****
2910 IF AL(I)<.1 THEN CM=UD(1) @ RETURN
2920 IF AL(I)>ALPHM THEN CM=UD(19) @ RETURN
2930 F=(LGT (AL(I))+1)/R49 @ Y=INT (F) @ Z=FP (F)
2940 CM=(1-Z)*UD(Y+1)+UD(Y+2)*Z @ RETURN
2950 !
2960 ! *************** PLOT DATA  ********************
2970 ! INPUTS : XL=MIN X, XR=MAX X, YT=MAX Y, YB=MIN Y
2980 LT=1 @ DEG
2990 AP=0 @ DISP "Add Plot";@ INPUT R$@ R$=UPC$ (R$)
3000 IF R$="Y" THEN AP=1 @ DISP "Enter LINE TYPE  1-8 " @ INPUT LT@ GOTO 3020
3010 IF R$#"N" THEN DISP "Y or N?" @ BEEP @ GOTO 2990
3020 IF PA=1 THEN LIMIT 0,125,0,75 @ X2=155 @ X3=85 @ GRAPH @ GOTO 3050
3030 IF IOFLAG THEN CONVERT OUT ZZZ ! Stop 0 Conv for ext plotter
3040 LIMIT 12,228,18,172 @ X2=120 @ X3=70
3050 IF AP=0 THEN GOSUB 3260
```

```
3060 LOCATE 18,X2,10,84 @ SETGU
3070 IF AP THEN 3130
3080 MOVE X3,89 @ CSIZE 5 @ LDIR 0 @ LORG 4 @ LABEL S$ @ CSIZE 4
3090 MOVE 18,87 @ LORG 3 @ LABEL TL$
3100 MOVE X2,87 @ LORG 9 @ LABEL TR$
3110 MOVE 18,5 @ LORG 3 @ LABEL LL$
3120 MOVE X2,5 @ LORG 9 @ LABEL DA$
3130 SCALE XL,XR,YB,YT
3140 MOVE XL+DXS,FF(1) @ I=1 @ LINE TYPE LT
3150 FOR X=XL+DXS+DX TO XR STEP DX
3160    I=I+1 @ DRAW X,FF(I)
3170 NEXT X
3180 IF PA=1 THEN 3200 ELSE PEN 0 @ MOVE XR,YT
3190 IF IOFLAG THEN CONVERT OUT ZZZ PAIRS ; ZZZ$@ RETURN ELSE RETURN
3200 ALPHA @ DISP "Dump to Printer";@ INPUT R$
3210 R$=UPC$ (R$) @ IF R$="N" THEN RETURN
3220 IF R$#"Y" THEN DISP "Y or N?" @ BEEP @ GOTO 3200
3230 IF PN=1 THEN GOSUB 7320 @ GOTO 3230 ! Select Printer(not CRT)
3240 IF PN=4 THEN DUMP GRAPHICS 0,0,0,-1 @ RETURN ELSE DUMP GRAPHICS @ RETURN
3250 !
3260 IF PA=1 THEN LIMIT 0,125,0,75 @ X2=155 @ X3=85 @ GRAPH @ GCLEAR @ FRAME
ELSE LIMIT 12,252,18,192 @ X2=134 @ X3=73
3270 LOCATE 18,X2,10,84 @ FRAME
3280 MOVE 4,50 @ LORG 4 @ LDIR 90 @ CSIZE 5,.7,30 @ LABEL YL$
3290 MOVE X3,94 @ CSIZE 8 @ LDIR 0 @ LABEL T$
3300 MOVE X3,5 @ LORG 6 @ CSIZE 5 @ LABEL X$
3310 NDX,NDY,YT,YB,I=0
3320 FOR X=XL+DXS TO XR STEP DX @ I=I+1 @ YT=MAX (YT,FF(I)) @ NEXT X
3330 YT=10*INT (YT/10)+10
3340 IF YT<= 10 THEN DELY=.2 @ GOTO 3380
3350 IF YT<= 25 THEN DELY=.5 @ NDY=1 @ GOTO 3380
3360 IF YT<= 50 THEN DELY=1 @ GOTO 3380
3370 IF YT<= 100 THEN DELY=2 ELSE DELY=4
3380 SCALE XL,XR,YB,YT @ CSIZE 4
3390 X=ABS (XL-XR) @ IF X<= 2.5 THEN DELX=.05 @ NDX=2 @ GOTO 3460
3400 IF X<= 5 THEN DELX=.1 @ NDX=1 @ GOTO 3460
3410 IF X<= 10 THEN DELX=.2 @ GOTO 3460
3420 IF X<= 25 THEN DELX=.5 @ NDX=1 @ GOTO 3460
3430 IF X<= 50 THEN DELX=1 @ GOTO 3460
3440 IF X<= 100 THEN DELX=2 @ GOTO 3460
3450 IF X<= 250 THEN DELX=4 ELSE DELX=10
3460 FXD NDX,NDY
3470 LGRID -DELX,DELY,XL,YB,5,5 @ RETURN
3480 !
3490 ! ******* CHANGE RADAR/SURFACE CLUT. PARAMETERS *******
3500 DISP "Select parameter to change"
3510 FOR I=1 TO NN
3520    IF I#13 THEN GOSUB 5730 ELSE GOSUB 5720
3530 NEXT I @ IF NN=9 THEN 3570
3540 I=31 @ GOSUB 5730 ! Tx El Bw
3550 I=32 @ GOSUB 5730 ! Elev GSL
3560 I=39 @ GOSUB 5730
```

```
3570 DISP "Enter # of parameter (0 to Quit, -1 to relist params.)";
3580 INPUT I@ IF I=0 THEN RETURN
3590 IF I=-1 THEN 3500
3600 IF NN=20 AND (I=31 OR I=32 OR I=39) THEN 3640
3610 IF I>NN THEN DISP "Use # from param list" @ BEEP @ GOTO 3500
3620 IF I=11 OR I=12 THEN GOSUB 3710 @ GOTO 3570
3630 IF I=13 THEN GOSUB 2860 @ GOSUB 3960 @ GOSUB 4010 @ GOTO 3570
3640 GOSUB 2870 @ IF I=14 OR I=19 THEN GOSUB 4010
3650 IF I<10 THEN GOSUB 3800
3660 IF I=32 THEN GOSUB 6090
3670 IF I=39 THEN GOSUB 7180
3680 GOTO 3570
3690 !
3700 ! ************ ENTER SURFACE CLUTTER MODEL *************
3710 DISP "Select ";M$(4) @ DISP
3720 FOR I=1 TO 3 @ DISP I;" ";M$(I) @ NEXT I @ DISP
3730 DISP "Enter Model #";@ INPUT CFLAG
3740 IF CFLAG<1 OR CFLAG>3 THEN DISP "USE # 1-3 ONLY" @ BEEP @ GOTO 3730
3750 IF CFLAG#CFLAGOLD THEN RROLD=0
3760 IF CFLAG=3 THEN GOSUB 4550 @ S$=M$(3)&" Clutter Model" @ RETURN
3770 I=10+CFLAG @ GOSUB 2870 @ CM=P(I) @ CFLAGOLD=CFLAG
3780 S$=M$(CFLAG)&"="&VAL$ (CM)&"dB" @ RETURN
3790 !
3800 ! *************** CALCULATE K1 ***************************
3810 X,K1=P(1)*P(2)*P(5)*P(5)/P(6)/P(8)*FNIPDB(P(3)+P(4)-ABS (P(7)))
3820 PR$=A$(23) @ GOTO 3860
3830 !
3840 ! *************** DISPLAY AND PRINT, TABULAR ********
3850 V$=UU$(RI) @ GOTO 3870
3860 V$=""
3870 DISP TAB (7);X;TAB (23);"= ";PR$;V$
3880 IF PN#1 THEN PRINT TAB (7);X;TAB (23);"= ";PR$;V$
3890 RETURN
3900 !
3910 ! ********************* DISPLAY, PRINT *************
3920 DISP PR$ @ IF PN#1 THEN PRINT PR$
3930 RETURN
3940 !
3950 ! ******* horizon for 4/3rds earth radius geometry *******
3960 HRAD=P(13)*UU(RI)/UU(HI) @ BB=(2*RE+HRAD)*HRAD
3970 X,RHOR=SQR (BB) @ AA=1+HRAD/RE @ DHOR=ACS (1/AA)
3980 PR$=A$(26) @ GOTO 3850
3990 !
4000 ! ************* duty cycle & no. of gates **********
4010 X,DUTY=P(14)*P(19)/1000000 @ FFLAG=0
4020 DELR=C2*UU(RI)/P(19) ! IPP in Range units
4030 IF DELR>RHOR OR RHORRN THEN 4040
4035 FFLAG=1 @ IF RHOR>100*DELR THEN DISP "PRF too high, reduce radar ht, or
PRF, or add RAM and redim. arrays DEP,FAC,AL" @ BEEP ! DIM 300 w/o HP Plotter
4040 PR$=A$(21) @ GOSUB 3860
4050 X,NG=INT (1/DUTY) @ PR$=A$(22) @ GOSUB 3860
4060 PR$="Ambig. range " @ X=DELR @ GOTO 3850
4070 !
```

```
4080 ! **************   Sweep S/N,Pd vs Range   ****************
4090 ! 2/2 DET, N=1,Pfa=3.87E-5,KCOR=3,Scan-Scan Swerling 1 Tgt. (8/12/83)
4100 DATA 30,1,1,.001,.003,.008,.019,.037,.066,.108,.164,.231,.306,.386,.466
4110 DATA .542,.613,.677,.732,.78,.821,.854,.882,.905,.924,.939,.951,.961
4120 DATA .969,.975,.98,.984,.987
4130 RESTORE 4100 @ READ NDET,SNMIN,DSN ! Use NDET<101
4140 FOR I=1 TO NDET @ READ PD(I)@ NEXT I
4150 GOSUB 4900 ! Enter RCS
4160 DISP "Enter max range and ";DR$;"range";UU$(RI);@ INPUT RMAX,DELR
4170 RR=DELR @ PR$=" RANGE "&UU$(RI)&"            S/N(dB)          Pd"
4180 GOSUB 3920 @ RZ=FNRM(SIGT,0)
4190 SN=FNPDB((RZ/RR)^4)
4200    F=1+(SN-SNMIN)/DSN @ IF F<= 1 THEN PD=PD(1) @ GOTO 4230
4210    IF F>= NDET THEN PD=PD(NDET) @ GOTO 4230
4220    I=INT (F) @ PD=FP (F)*(PD(I+1)-PD(I))+PD(I)
4230    DISP RR,SN,PD @ IF PN#1 THEN PRINT TAB (5);RR,SN,PD
4240    RR=RR+DELR
4250 IF RR<= RMAX THEN 4190
4260 DISP "NEW RCS";@ INPUT R$@ R$=UPC$ (R$) @ IF R$="Y" THEN 4150
4270 IF R$#"N" THEN DISP "Y or N?" @ BEEP @ GOTO 4260 ELSE 6290
4280 !
4290 ! **********   SURFACE C/N VS RANGE SWEEP   *************
4300 DISP "Unambiguous PRF, enter ";DR$;"range";UU$(RI);"for sweep";
4310 INPUT DELR@ I,J=1 @ NP,NT=0
4315 IF RHOR/DELR>301 THEN DISP "Use ";DR$;"R>";RHOR/301 @ BEEP @ GOTO 4300
4320 FOR RR=DELR TO RHOR STEP DELR
4330    PR$="Range" @ X=RR @ GOSUB 3850 ! print range
4340    GOSUB 2560 @ GOSUB 2760 ! calc Dep, Graz angles, FAC, times gain
4350    SUMC=X @ GOSUB 2660 @ FF(J)=CTOUT @ J=J+1 ! C/N (dB)
4360    IF SFLAG=4 THEN GOSUB 4940 @ PR$="S/N(dB)="&VAL$ (SNDB)&",   S/(C+N)
(dB)="&VAL$ (SNCDB) @ GOSUB 3920 @ GOTO 4380
4370    IF SFLAG=3 THEN SIGT=1 @ GOSUB 4940 @ PR$="Min RCS="&VAL$
(FNIPDB(TNDB-SNCDB))&" sq. meters" @ GOSUB 3920
4380 NEXT RR
4390 DISP "PLOT";@ INPUT R$@ R$=UPC$ (R$) @ IF R$="N" THEN 1810
4400 IF R$#"Y" THEN DISP "Y or N?" @ BEEP @ GOTO 4390
4410 XL=0 @ XR=RHOR @ DX,DXS=DELR @ X$="Range "&UU$(RI)
4420 TR$="Tx Elev="&VAL$ (P(16))&"deg" @ TL$="Rx Elev="&VAL$ (P(17))&"deg"
4430 LL$="Alt"&UU$(HI)&VA$(P(13)) @ GOSUB 2970 @ GOTO 1810 ! Plot
4440 !
4450 ! **************   Select User-Defined Clutter Model   **************
4460 DISP "Select clutter Model"
4470 DISP "1   New User-defined model"
4480 DISP "2   Model:Ex. 4-1.2, Radar Calcs..."
4490 DISP "3   Other Models can be added"
4500 DISP @ DISP "Enter Model #";@ INPUT R
4510 IF R<1 OR R>3 THEN DISP "Use # 1-3" @ BEEP @ GOTO 4500
4520 ON R GOTO 4540,4760,4800
4530 !
```

```
4540 ! ********  USER-DEFINED SURFACE CLUTTER MODEL  ********************
4550 DISP "Enter max grazing angle (deg)";@ INPUT X@ ALPHM=X
4560 IF X>90 OR X<= 0 THEN DISP "Use  0< ANGLE <=90 " @ BEEP @ GOTO 4550
4570 R49=(1+LGT (X))/18
4580 R3=10^R49 @ R1=.1
4590 IF PN#1 THEN PRINT " GRAZ(DEG)        SIGMA-ZERO(dB)"
4600 FOR I=1 TO 19
4610    DISP "Enter Sigma-zero(";R1;" deg)in dB";@ INPUT UD(I)
4620    IF PN#1 THEN PRINT R1,UD(I) @ R1=R1*R3
4630 NEXT I
4640 DISP "Corrections";@ INPUT R$@ R$=UPC$ (R$) @ IF R$="N" THEN RETURN
4650  IF R$#"Y" THEN DISP "Y or N?" @ BEEP @ GOTO 4640
4660  DISP "Enter grazing angle(deg) and sigma-zero (dB)";@ INPUT R1,UDR1
4670  IF R1>= .1 AND R1<= X THEN 4690
4680 DISP "Use 0.1<=grazing angle<=";X @ BEEP @ GOTO 4660
4690 I=LGT (10*R1)/R49+1 @ UD(I)=UDR1 @ GOTO 4640
4700 !
4710 ! *********** SUB TO READ CLUTTER MODEL  *****************
4720 READ ALPHM@ R49=(1+LGT (ALPHM))/18
4730 FOR I=1 TO 19 @ READ UD(I)@ NEXT I @ READ PR$@ GOTO 3920
4740 !
4750 ! **********  clutter Model Ex. 4-1.2 Radar Calc.  ***************
4760 RESTORE 4770 @ GOTO 4710 ! Sea clut., 13' waves, upwind, S-band
4770 DATA 15,-65,-61,-58,-55,-53,-51,-49,-48,-48,-47,-45,-44,-44,-43,-42
4780 DATA -41,-40,-38.5,-37,Sea Clutter; 13 ft waves; upwind; S-band
4790 !
4800 !  Place-holder to illustrate format:(R3=10^R49)
4810 ! DATA ALPHM, UD(.1), UD(.1*R3),...,UD(.1*R3^18), PR$ (model description)
4820 RETURN
4830 !
4840 ! *************  Enter  T/N AND TARGET LOSSES  ***************
4850 PR$="S/N (dB) for detection " @ DISP "Enter ";PR$;@ INPUT X@ TNDB=X
4860 V$="" @ GOSUB 3880 @ I=40 @ GOSUB 2860
4870 K1T=K1*FNIPDB(ABS (P(7))-ABS (P(40))) @ RETURN
4880 !
4890 ! ****************** Enter  RCS  ************************
4900 PR$="RCS(sq. meters)" @ DISP "Enter ";PR$;@ INPUT X
4910 IF X<= 0 THEN DISP "Use RCS >0" @ BEEP @ GOTO 4900
4920 SIGT=X @ GOTO 4860
4930 !
4940 ! ************  calc S/N, S/(N+C) for C/N vs Range Sweep  ****
4950 SNDB=24.92+P(9)+FNPDB(X*K1T*SIGT*UU(RI)/RR/F)
4960 SNCDB=SNDB-FNPDB(1+FNIPDB(CTOUT)) @ RETURN
4970 !
4980 ! ****************** RAIN CLUTTER  ********************
4990 T$="**RAIN/CHAFF CLUTTER**" @ I=10 @ GOSUB 2870 ! Azimuth beamwidth
5000 I=13 @ GOSUB 2860 @ GOSUB 3960 ! Enter RH, Calc AA,BB,RHOR
5010 I=29 @ GOSUB 2860 @ GOSUB 6060 ! Rain height
5020 DISP "Select type of entry"
5030 DISP "1  rainfall rate"
5040 DISP "2  chaff coef"
5050 DISP "Enter # ";@ INPUT R@ RVMIN=0
```

```
5060 IF R<1 OR R>2 THEN DISP "1 or 2?" @ BEEP @ GOTO 5050
5070 I=26+R @ GOSUB 2870 @ IF R=1 THEN S$="Rain rate="&VAL$ (P(27))&"mm/hr" @
GOSUB 6030 ELSE S$="RCS="&VAL$ (P(28))&"sqm/cum" @ GOSUB 5220 ! Clut entry
5080 I=15 @ GOSUB 2870 ! Elevation beamwidth
5090 I=39 @ GOSUB 2870 @ GOSUB 7180 ! Ant sep & beam xover range
5100 I=30 @ GOSUB 2870 @ GOSUB 6000 ! Beam Elev
5110 I=14 @ GOSUB 2870 ! Pulse width
5120 I=19 @ GOSUB 2870 ! PRF
5130 GOSUB 4010 ! Duty cycle, no. gates, DELR
5140 DISP T$;" Menu: Select next operation" @ DISP
5150 DISP "1  New parameters"
5160 DISP "2  Calc Rain Clutter "
5170 DISP "3  Main Menu"
5180 DISP @ DISP "Enter #";@ INPUT R
5190 IF R<1 OR R>3 THEN DISP "Use # 1-3" @ BEEP @ GOTO 5180
5200 ON R GOTO 5280,5300,1560
5210 !
5220 ! ******************* Enter min chaff range  ****************
5230 Y=RHOR+RHORRN
5240 PR$="Min chaff range"&UU$(RI) @ DISP "Enter ";PR$;"0<=R<=";Y;@ INPUT X
5250 IF X<0 OR X>Y THEN DISP "Use 0<=R<=";Y @ BEEP @ GOTO 5240
5260 RVMIN=X @ GOTO 4860
5270 !
5280 GOSUB 5490 ! Change parameters, calc rain clut
5290 !
5300 ! ******************* CALC RAIN CLUTTER   ****************
5310 DISP "Caution! Array FF may overflow if ";DR$;"R is too small"
5315 DISP "Enter ";DR$;"Range ";UU$(RI);"(<=";DELR;")" @ INPUT DRR
5320 IF DRR>DELR THEN DISP "Use ";DR$;"R < ";DELR @ GOTO 5310
5330 RRS=DRR @ F4,SUMOUT=0 @ R25=38160000*K1*P(10)*P(28) @ J=1
5340 PR$="" @ GOSUB 3920 @ X$="Range "&UU$(RI)
5350 PR$="    "&X$&"      Output C/N (dB)" @ GOSUB 3920
5360 GOSUB 5770 @ IF F1 THEN 5410 ELSE FF(J)=CTOUT @ J=J+1
5370    SUMOUT=SUMOUT+FNIPDB(CTOUT)
5380    RRS=RRS+DRR
5390 IF SUMC THEN F4=1 @ GOTO 5360
5400 IF F4=0 THEN 5360
5410 X=FNPDB(SUMOUT/J)-P(9) ! C/N at IF for CW only
5420 PR$="C/N,IF,CW (dB)" @ GOSUB 3860
5430 DISP "Plot";@ INPUT R$@ R$=UPC$ (R$) @ IF R$="N" THEN 5140
5440 IF R$#"Y" THEN DISP "Y or N?" @ BEEP @ GOTO 5430
5450 XL=0 @ XR=MIN (DELR,RHOR+RHORRN) @ DX,DXS=DRR
5460 LL$="Elev="&VAL$ (P(30))&"deg" @ TR$,TL$="" @ GOSUB 2970 @ GOTO 5140
5470 !
5480 ! ************ CHANGE RAIN PARAMETERS  ******************
5490 DISP "Select rain parameter to change"
5500 FOR I=1 TO 10 @ GOSUB 5730 @ NEXT I
5510 I=13 @ GOSUB 5720
5520 FOR I=14 TO 15 @ GOSUB 5730 @ NEXT I
5530 I=19 @ GOSUB 5730
5540 FOR I=27 TO 30 @ IF I=29 THEN GOSUB 5720 ELSE GOSUB 5730
```

```
5550 NEXT I @ I=39 @ V$="(ft)" @ GOSUB 5740
5560 DISP "Enter # of parameter (0 to Quit, -1 to relist params)";
5570 INPUT I@ IF I=0 THEN RETURN
5580 IF I=-1 THEN 5490 ! relist
5590 IF I=13 OR I=29 THEN GOSUB 2860 ELSE GOSUB 2870 ! Enter param
5600 IF I=5 THEN GOSUB 6030
5610 IF I<10 THEN GOSUB 3800 @ GOTO 5560
5620 IF I=13 THEN GOSUB 3960 @ GOSUB 4010 @ GOTO 5560
5630 IF I=14 OR I=19 THEN GOSUB 4010 @ GOTO 5560
5640 IF I=15 OR I=30 THEN GOSUB 6000 @ GOTO 5560
5650 IF I=27 THEN GOSUB 6030 @ GOTO 5560
5660 IF I=28 THEN GOSUB 5220 @ GOTO 5560
5670 IF I=29 THEN GOSUB 6060 @ GOTO 5560
5680 IF I=39 THEN GOSUB 7180 ELSE DISP "Not a rain clut. param.!" @ BEEP
5690 GOTO 5560
5700 !
5710 ! ********* sub to display parameters  *********
5720 V$=UU$(HI) @ GOTO 5740
5730 V$=""
5740 DISP I;"    ";A$(I);V$;"=";P(I) @ RETURN
5750 !
5760 ! ********* SUB FOR SINGLE RANGE RAIN CLUT  *********
5770 DELD,F1,F5,SUMC=0 @ RR=RRS
5780 IF RR>DELR THEN F1=1 @ RETURN
5790 IF HRAD<HRAIN THEN F5=1 @ GOTO 5880
5800 DTRN=ACS ((1+HRAIN/RE)/AA)
5810 IF DTRN>DBB THEN F1=1 @ RETURN ELSE 5880
5820 !
5830 F5=1 @ Y=UU(RI)/RR @ IF RR<= RC OR RR<RVMIN THEN 5870
5840 IF RR>10*RC THEN 5860
5850 X=RC/RR @ DELD=DELD*(ACS (X)/90-2*X*SQR (1-X*X)/PI ) ! Tx,Rx Overlap
5860 SUMC=SUMC+DELD*Y*Y
5870   RR=RR+DELR
5880   IF RR>RHOR+RHORRN THEN 5940 ! Exit if beyond horizon
5890   IF RR>RHOR THEN DD=DHOR ELSE H=0 @ GOSUB 1420 @ DD=DEP
5900   H=HRAIN @ GOSUB 1420
5910   DELD=MIN (DD,DBB)-MAX (DEP,DTB) ! Eff Beamwidth
5920 IF DELD>0 THEN 5830
5930 IF F5=0 THEN 5870
5940 CTOUT=FNPDB(SUMC*R25*P(14))+P(9)
5950 DISP USING F1 ; RRS,CTOUT @ IF PN#1 THEN PRINT USING F1 ; RRS,CTOUT
5960 F1: IMAGE 5D.3D,10X,4D.3D
5970 RETURN
5980 !
5990 ! ************* DEP OF TOP AND BOTTOM OF BEAM  **************
6000 DTB=-(P(30))-P(15)/2 @ DBB=DTB+P(15) @ RETURN ! Beam top,bottom
6010 !
6020 ! ****************** RAIN COEF FROM RAIN RATE  **************
6030 X,P(28)=.00000569*P(27)^1.6/P(5)^4 @ PR$=A$(28) @ GOTO 3860
6040 !
```

```
6050 ! ******************** RAIN  HT  & HORIZ RANGE  **********
6060 HRAIN=P(29)*UU(RI)/UU(HI) @ RHORRN=SQR (2*RE+HRAIN)*HRAIN @ RETURN
6070 !
6080 ! **************** ELEV SIDELOBE LEVEL  calcs ***************
6090 GSLDB=ABS (P(32))
6100 XMAX=SQR (GSLDB/40/LGT (2)) @ GSL=FNIPDB(-GSLDB) @ RETURN
6110 !
6120 NN=9 @ GOSUB 3500
6130 !
6140 ! ******************** S/N, and J/N  *********************
6150 GOSUB 4850 ! Enter T/N,tgt loss
6160 PR$="" @ GOSUB 3920 @ DISP " ** S/N, J/N Menu ** Select Next Operation"
6170 DISP @ DISP "1  New Radar Parameters"
6180 DISP "2  Clear Range Performance"
6190 DISP "3  ECM Range Performance "
6200 DISP "4  Alpha Range"
6210 DISP "5  Main Menu" @ DISP
6220 DISP "Enter #";@ INPUT R
6230 IF R<1 OR R>5 THEN DISP "Use # 1-5" @ BEEP @ GOTO 6220
6240 ON R GOTO 6120,6290,6540,6810,1560
6250 !
6260 NN=9 @ GOSUB 3500 @ GOSUB 4850
6270 !
6280 ! ***************** Clear range Perf  ************
6290 PR$="" @ GOSUB 3920 @ DISP "Select Clear Perf Option"
6300 DISP "1  New Radar Parameters"
6310 DISP "2  Sweep S/N and Pd vs. range"
6320 DISP "3  Sweep Detection Range vs. RCS"
6330 DISP "4  Ro for RCS (S/N=1)"
6340 DISP "5  S/N, J/N Menu"
6350 DISP @ DISP "Enter #";@ INPUT R
6360 IF R<1 OR R>5 THEN DISP "Use # 1-5" @ BEEP @ GOTO 6350
6370 ON R GOTO 6260,4080,6390,6500,6160
6380 !
6390 ! ************ Sweep Range vs. RCS  ******************
6400 RTM1=FNRM(1,TNDB) ! Calc  range for 1 sq. m.
6410 DISP "Enter min, max and ";DR$;"RCS (sq. m.)";
6420 INPUT SIGMIN,SIGMAX,DELSIG
6430 IF SIGMIN<= 0 OR SIGMAX<= 0 OR DELSIG<= 0 THEN DISP "Use values > 0" @
BEEP @ GOTO 6410
6440 PR$="    RCS (sq. m.)        RANGE ("&UU$(RI)&")" @ GOSUB 3920
6450 FOR SIGT=SIGMIN TO SIGMAX STEP DELSIG
6460   RR=RTM1*SIGT^.25
6470   DISP SIGT,RR @ IF PN#1 THEN PRINT TAB (5);SIGT,RR
6480 NEXT SIGT @ GOTO 6290
6490 !
6500 ! ***************** Ro for RCS  ********************
6510 GOSUB 4900 @ X=FNRM(SIGT,0) @ PR$="Ro, Marcum's Idealized Range"
6520 GOSUB 3850 @ GOTO 6290
6530 !
```

```
6540 ! *****************   ECM RANGE PERF.   ****************
6550 FOR I=33 TO 35 @ GOSUB 2870 @ NEXT I ! LJ,ERPD,BWJ
6560 I=37 @ GOSUB 2850 ! Standoff jammer range
6570 BE=MIN (P(8),P(35)) ! Effective Bandwidth
6580 C=FNCC(P(34))*BE/P(8)
6590 GOSUB 4900 ! Enter RCS
6600 X,RTM=FNRM(SIGT,TNDB) @ M$="Monostatic" @ B$="Bistatic" ! Mono range
6610 PR$="" @ GOSUB 3920 @ DISP "Select Radar Type" @ DISP
6620 DISP "1 ";M$
6630 DISP "2 ";B$ @ DISP
6640 DISP "Enter #";@ INPUT R
6650 IF R<1 OR R>2 THEN DISP "Use 1 or 2" @ BEEP @ GOTO 6640
6660 RFLAG=R-1 @ P$=M$
6670 IF R=2 THEN I=38 @ GOSUB 2850 @ X,RTB=RTM*RTM/P(38) @ P$=B$
6680 PR$="Clear Range, "&P$ @ GOSUB 3850 @ DISP "Select ant. gain entry method"
6690 DISP @ DISP "1  Sweep Jammer-to-target az separation"
6700 DISP "2  Input radar Rx gain in jammer direction"
6710 DISP @ DISP "Enter #";@ INPUT R@ IF R=1 THEN 7090
6720 IF R#2 THEN DISP "Use 1 or 2" @ BEEP @ GOTO 6710
6730 !
6740 ! ******************   Radar Gain in Jammer Direction   **********
6750 PP$="Rx Gain WRT Peak  (dB)" @ DISP "Enter ";PP$;"in jammer direction ";
6760 DISP "([EL] to Quit";@ INPUT R$@ IF R$="" THEN 6160
6770 GRJDB=VAL (R$) @ GRJ=FNIPDB(GRJDB)
6780 IF GRJDB>0 THEN DISP "Gain (dB) must be <=0" @ BEEP @ GOTO 6750
6790 PP$=PP$&"="&VAL$ (GRJDB) @ GOSUB 6960 @ GOTO 6750
6800 !
6810 ! ********************   ALPHA RANGE   *****************
6820 I=33 @ GOSUB 2870 @ C=FNCC(.001) @ RTM=FNRM(1,TNDB)
6830 X=SQR (SQR (C*C+RTM^4)-C)
6840 PR$="Alpha range" @ GOSUB 3850 @ GOTO 6160
6850 !
6860 ! ********   Select Range and Alt Units   *******
6870 DISP @ DISP "Select Range, Alt Units" @ DISP
6880 DISP "1   km "
6890 DISP "2   nmi"
6900 DISP "3   mi (statute)"
6910 DISP "4   ft" @ DISP
6920 DISP "Enter #s for Range and alt (ie 1,4 for km,ft)";@ INPUT RI,HI
6930 IF RI<1 OR RI>4 OR HI<1 OR HI>4 THEN DISP "Use #s 1-4" @ BEEP @ GOTO 6920
6940 RE=4588*UU(RI) @ RETURN
6950 !
6960 ! **************   SUB FOR STANDOFF AND ESCORT JAMMER RANGE PERF   ******
6970 CJ=C*GRJ @ JN=2*CJ/P(37)/P(37)
6980 IF RFLAG THEN 7040 ! Branch for Bistatic
6990 RSO=RTM/(1+JN)^.25 @ CT=SQR (CJ)/100 ! Monostatic
7000 IF RTM<CT THEN RES=RTM*RTM/SQR (2*CJ) ELSE RES=SQR (SQR (CJ*CJ+RTM^4)-CJ)
7010 PR$=PP$&",  RSO="&VAL$ (FNRD(RSO))&UU$(RI)&",  RES="&VAL$ (FNRD(RES))
7020 PR$=PR$&UU$(RI)&",  J/N(dB)="&VAL$ (FNRD(FNPDB(JN))) @ GOTO 3920
7030 !
```

```
7040 RSO=RTB/SQR (1+JN) ! Bistatic
7050 X=RTB*RTB-2*CJ @ RES=0
7060 IF X>0 THEN RES=SQR (X)
7070 GOTO 7010
7080 !
7090 ! **********  JAMMER-TO-TGT AZ SEP SWEEP  ***********
7100 I=10 @ GOSUB 2870 ! AZ BW
7110 I=36 @ GOSUB 2870 @ GSLDB=ABS (P(36)) @ GOSUB 6100 ! GSL-AZ
7120 PR$="" @ GOSUB 3920
7130 FOR AZ=0 TO 2.5*P(10) STEP P(10)/10 @ th=AZ @ bw=P(10)
7140    GOSUB 1380 @ GRJ=GAIN ! Gain WRT peak, Gaussian shape
7150    PP$="Az sep (deg)="&VAL$ (AZ) @ GOSUB 6960 @ IF JN<.02 THEN 6160
7160 NEXT AZ @ GOTO 6160
7170 !
7180 ! *******  SUB TO ENTER ANTENNA SEP, CALC BM XOVER RANGE ********
7190 IF P(39)=0 THEN RC=0 @ RETURN
7200 DISP "Enter V or H for separation direction";@ INPUT SEP$
7210 SEP$=UPC$ (SEP$) @ IF SEP$="V" THEN I=15 @ PR$="Vertical" @ GOTO 7240
7220 IF SEP$="H" THEN I=10 @ PR$="Horizontal" @ GOTO 7240
7230 DISP "Use V or H" @ BEEP @ GOTO 7190
7240 RC=P(39)*180/PI /P(I)*UU(RI)/UU(4) ! crossover range in range units
7250 IF PN#1 THEN PRINT TAB (25);PR$;" Dimension"
7260 RETURN
7270 !
7280 ! ******  INITIALIZATION &  PRINTER SELECTION  *****
7290 PAGESIZE 24 @ CLEAR @ DEG @ PRINTER IS 1 @ IOFLAG=0
7300 DISP TAB (10);T$ @ DISP
7310 DISP "Enter today's date (any comma-less format)";@ INPUT DA$
7320 DISP "Select Printer:" @ DISP
7330 DISP "1   CRT"
7340 DISP "2   HP-82905B--HP-IB"
7350 DISP "3   HP-82905B--Centronics"
7360 DISP "4   MX-100" @ DISP
7370 DISP "Enter #";@ INPUT PN
7380 IF PN<1 OR PN>4 THEN DISP "Use # 1-4 " @ BEEP @ GOTO 7370
7385 IF PN#1 THEN DISP "Make sure Top-Of-Form is set properly. Press [END
LINE];" @ INPUT PR$
7390 ON PN GOTO 7490,7410,7420,7460
7400 !
7410 PRINTER IS 301 @ ZZZ=3 @ GOTO 7430
7420 PRINTER IS 701 @ ZZZ=7
7430 ON ERROR GOTO 7450 @ a=BTD ("1") ! I/O ROM Test
7440 CONVERT OUT ZZZ PAIRS ; ZZZ$@ IOFLAG=1 ! Normal to slashed 0 conv
7450 OFF ERROR @ PRINT CHR$ (27)&"&l1L" @ GOTO 7480
7460 PRINTER IS 701 @ PRINT CHR$ (27);"N";CHR$ (8)
7480 PRINT TAB (15);T$ @ PRINT TAB (25);DA$
7490 DISP @ DISP "Answer questions with 'Y' or 'N', [END LINE]" @ DISP
7500 RETURN
7510 !
```

```
7520 ! ****** Sub to Select plotter, load binary GDUMP ****
7530 DISP "Enter Plotter Address (CRT=1)";@ INPUT PA
7540 IF PA<1 THEN DISP "Use # >=1" @ BEEP @ GOTO 7530
7550 ON ERROR GOTO 7570 @ GDUMP @ GOTO 7620 ! Error if GDUMP not resident,
7560 ! displays "Copyright Hewlett-Packard 203.09", if resident
7570 ON ERROR GOTO 7600 @ LOADBIN "GDUMP" @ GOTO 7550 ! several errors possible
7580 ! ERRN=67 if GDUMP not on disc
7590 ! ERRN=130 if bad disc, no disc, door open, etc.
7600 IF ERRN =67 OR ERRN =130 THEN DISP "Insert disc with binary GDUMP, press
[CONT]" @ PAUSE @ GOTO 7570
7610 DISP "Unknown error,ERRN=";ERRN ;" try again" @ GOTO 7530
7620 OFF ERROR @ PLOTTER IS PA @ RETURN
```

Chapter 5 Filters and Filtering

Program DIGFILT

```
10 ! Pgm DIGFILT, Rev 2/28/84
20 ! Translated from Pgms 5-1,2 and 3 of Radar Calc...by W. Skillman
30 !
40 ! Improvements vs Radar Calc.
50 !    Response plot: Plotter or CRT with dump to printer
60 !    Zeros and Pole locations calc on Coef entry
70 !    Clutter Attn, Imp Factor for any filter
75 !    Noise gain each output for impulse response
80 !
90 ! Important Variables:
100 !     AM.....Ampl of freq. response (voltage)
110 !     CA.....Clutter Attenuation (power)
120 !     C(I,J).Coef. array, I=coef. index, J=pair index
130 !     Ci.....Dbl delay line coefs, i=0-3
140 !     CZ.....Sngl Zero Coef.
150 !     FF.....array for freq. resp plot (dB)
160 !     FLAG1..0=N even, 1=N odd
170 !     FT.....Normalized freq
180 !     F3T....-3dB bandwidth for Butterworth and Chebyshev filters
190 !     GA.....Filter pwr gain
200 !     GAVG...Avg pwr gain
210 !     GREF...dB ref for freq. resp
220 !     IFLAG..0=Impulse, 1=Step Resp
230 !     M......1=High Pass, -1=Low Pass, Cheby or Butter
240 !     N......# of delay lines
250 !     N2.....INT(N/2)
260 !     OM.....scaling factor for Butter & Cheby filters
270 !     PH.....Phase of filter freq. response (deg)
280 !     R......Cheby filter ripple (dB)
285 !     REF....Ref gain for dB (dB)
290 !     RP.....Single Pole coef. or location in Z-plane
300 !     RZ.....Single Zero location
310 !     T(I,J).Delay line outputs,(voltage) I=1 or 2, J= pair index
315 !     TF.....1=User-Def, 2=Butter, 3=Cheby
320 !     TODD...Single delay output
330 !
340 DIM C(4,25),T(2,25),T$[30],PR$[99],X$[35],YL$[35],S$[35],FF(201)
350 DIM SZ$[25],DD$[25]
360 !
370 ZZZ$="0"&CHR$ (210) @ SZ$="Single Zero/Pole:  C0=" @ BW$=" -3dB Bandwidth"
380 DD$=" double delay line pair " @ CC$="Chebyshev" @ ZP$="Zero/Pole"
390 T$="**DIGITAL FILTER ANALYSIS**" @ UP$="Unstable pole!"
395 CB$=" clutter bandwidth" @ BF$="Butterworth " @ GOTO 610
400 !
410 ! *************** Defined Functions ********************
420 DEF FNVDB(AM) = 20*LGT (MAX (EPS ,ABS (AM))) ! Voltage ratio to dB
430 DEF FNPDB(P) = 10*LGT (MAX (EPS ,P)) ! Power ratio to dB
440 DEF FNPATN2(Y,X) = ATN2 (Y,X+EPS ) ! Arctan protected from 0,0 error
450 DEF FNRD(X) = INT (1000*X+.5)/1000 ! Round to 3 decimals
```

```
460 DEF FNCOSH(X) = (EXP (X)+EXP (-X))/2 ! Hyperbolic cosine
470 DEF FNICOSH(X) = LOG (ABS (X)+SQR (X*X-1)) ! Inv Hyp cos
480 DEF FNPTAN(X) = TAN (X+.0000000001) ! TAN Prot at odd multiples of 90 deg
490 !
500 ! ********* Sub to find quadratic roots *******************
510 ! X1, X2 are sol'ns to X^2+bX+c=0
520 CFLAG=0 @ X1=-(b/2) @ bc=X1*X1-c @ X2=SQR (ABS (bc))
530 IF bc>0 THEN X1=X1+SGN (X1)*X2 @ X2=c/X1 @ GOTO 550 ! Real, unequal roots
540 IF bc THEN CFLAG=1 ELSE X2=X1 ! Complex pair or real, equal roots
550 X1$=VAL$ (X1) @ X2$=VAL$ (X2) @ RETURN
560 !
570 ! ********** Sub to calc. Chebyshev Polynomial **********
580 IF ABS (X)>1 THEN TN=FNCOSH(N*FNICOSH(X)) ELSE TN=COS (N*ACS (X))
590 RETURN
600 !
610 GOSUB 2940 @ GOSUB 3480 ! init, printer sel.,load GDUMP, sel. plotter
620 DISP "Note: this program handles only REAL coefficients" @ DISP
630 GOSUB 770 ! filter data entry
640 !
650 ! ************ Main Menu ********************************
660 PR$="" @ GOSUB 1290 @ DISP "Main Menu: select next operation"
670 DISP "1  New filter"
680 DISP "2  Frequency response"
690 DISP "3  Impulse response"
700 DISP "4  Step response"
710 DISP "5  Clutter Attenuation/Gavg/Imp Fact" @ DISP "6  Disc Menu"
720 DISP @ DISP "Enter # ";@ INPUT R@ IF R=6 THEN CHAIN "Autost"
730 IF R<1 OR R>6 THEN DISP "Use # 1-6" @ BEEP @ GOTO 720
740 ON R GOSUB 770,1960,2370,2370,2560 @ GOTO 660
750 !
760 ! ************ Enter Filter Data *************************
770 DISP "Enter # of delay lines in filter (1-25)";@ INPUT N
780 IF N<1 OR N>25 THEN DISP "Use # 1-25" @ BEEP @ GOTO 770
790 IF FP (N) THEN DISP "Use integer" @ BEEP @ GOTO 770
800 PR$="Filter config. is shown in Fig" @ N2=INT (N/2)
810 FLAG1=1 @ IF FP (N/2) THEN FLAG1=0 @ PR$=PR$&" 5-1.1" @ GOTO 830
820 IF N=1 THEN PR$=PR$&" 5-1.2," ELSE PR$=PR$&"s 5-1.1,5-1.2"
830 PR$=PR$&" for a "&VAL$ (N)&" delay line filter" @ GOSUB 1290
840 !
850 ! ************* Filter Type Menu ********************
860 PR$="" @ GOSUB 1290 @ DISP "Select Filter Type:" @ DISP
870 DISP "1  User-Defined"
880 DISP "2  ";BF$
890 DISP "3  ";CC$ @ DISP
900 DISP "Enter #";@ INPUT TF
910 IF TF<1 OR TF>3 THEN DISP "Use # 1-3" @ BEEP @ GOTO 900
920 ON TF GOTO 940,1590,1780
930 !
940 ! ******* User-Defined Filter Data Entry **************
950 PR$="User-defined Filter" @ GOSUB 1290
955 DISP "Select type of data entry: " @ DISP
```

```
960 DISP "1  Filter Coefficients"
970 DISP "2  ";ZP$;" locations in Z-plane"
980 DISP @ DISP "Enter no. ";@ INPUT R@ IF R=1 THEN 1320
990 IF R#2 THEN DISP "Use 1 or 2" @ BEEP @ GOTO 980
1010 !
1020 ! ************* Entry of Zero/Pole Locations ******************
1030 IF N=1 THEN 1240
1040 DISP "Enter ";ZP$;" locations in pairs, separated by a comma"
1050 DISP "  If the pair is complex conjugate, place 'j', or 'J'  before "
1060 DISP "the second or Imaginary value..ie:  4.35,j6.54"
1070 FOR I=1 TO N2 ! Enter zeros and poles for double delay pairs
1080    PR$=ZP$&" Pair #"&VAL$ (I) @ IF PN#1 THEN PRINT PR$
1090    DISP "Enter Zero pair #";I;@ INPUT R,Z$@ Z$=UPC$ (Z$)
1100    IF POS (Z$,"J") THEN IM=VAL (Z$[2]) @ GOTO 1120
1110    R2=VAL (Z$) @ PR$=", Real="&Z$ @ C0=R*R2 @ C3=-R-R2 @ GOTO 1130
1120    PR$=", Imag="&Z$[2] @ C0=R*R+IM*IM @ C3=-(2*R)
1130    PR$=" Zero: Real="&VAL$ (R)&PR$&", C0="&VAL$ (C0)&", C3="
1140    PR$=PR$&VAL$ (C3) @ GOSUB 1290
1150    DISP "Enter Pole pair #";I;@ INPUT R,Z$@ Z$=UPC$ (Z$)
1160    IF POS (Z$,"J") THEN IM=VAL (Z$[2]) @ GOTO 1190
1170    R2=VAL (Z$) @ PR$=", Real="&Z$ @ C1=-(R*R2) @ C2=R+R2
1180    IF ABS (R)>1 OR ABS (R2)>1 THEN DISP UP$ @ BEEP @ GOTO 1150 ELSE 1210
1190    X=R*R+IM*IM @ IF X>1 THEN DISP UP$ @ BEEP @ GOTO 1150 !  Pole test
1200    PR$=",  Imag="&Z$[2] @ C1=-X @ C2=2*R
1210    PR$=" Pole: Real="&VAL$ (R)&PR$&", C1="&VAL$ (C1)&", C2="
1220    PR$=PR$&VAL$ (C2) @ GOSUB 1290 @ GOSUB 1570
1230 NEXT I @ IF FLAG1 THEN RETURN
1240 DISP "Enter Single Zero, Pole";@ INPUT RZ,RP@ CZ=-RZ
1250 IF ABS (RP)>1 THEN DISP UP$ @ BEEP @ GOTO 1240
1255 PR$=SZ$&VAL$ (CZ)
1260 PR$=PR$&", C1="&VAL$ (RP)&",  Zero="&VAL$ (RZ)&", Pole="&VAL$ (RP)
1270 !
1280 ! ************* disp/print ********************************
1290 DISP PR$ @ IF PN#1 THEN PRINT PR$
1300 RETURN
1310 !
1320 ! *************** Entry of Filter Coefficients ***************
1330 IF N=1 THEN 1400
1340 DISP "Enter 4 filter coefficients for each";DD$
1350 IF FLAG1=0 THEN DISP "  then 2 coefs for single delay line."
1360 FOR I=1 TO N2
1370    DISP "Enter C0,C1,C2,C3 for";DD$;" # ";I;@ INPUT C0,C1,C2,C3
1380    FF=0 @ GOSUB 1450 @ IF FF THEN 1370
1390 NEXT I @ IF FLAG1 THEN RETURN
1400 DISP "Enter C0,C1 for single delay line";@ INPUT CZ,RP
1410 IF ABS (RP)>1 THEN DISP UP$ @ BEEP @ GOTO 1400 ELSE RZ=-CZ @ GOTO 1255
1430 !
1450 ! ********* Sub to calc Zeros/Poles from coefs, disp/print *******
1480 b=-C2 @ c=-C1 @ GOSUB 520 @ IF CFLAG THEN 1510
1490 IF ABS (X1)>1 OR ABS (X2)>1 THEN DISP UP$ @ BEEP @ FF=1 @ RETURN
1500 GOSUB 1564 @ PR$=" Real Poles: Re1="&X1$&", Re2="&X2$ @ GOTO 1530
1510 IF X1*X1+X2*X2>1 THEN DISP UP$ @ BEEP @ FF=1 @ RETURN
1520 GOSUB 1564 @ PR$=" Complex Pole Pair: Re="&X1$&", Im="&X2$
```

```
1530 GOSUB 1290 @ b=C3 @ c=C0 @ GOSUB 520 @ IF CFLAG THEN 1550
1540 PR$="   Real Zeros: Re1="&X1$&", Re2="&X2$ @ GOTO 1290
1550 PR$="   Complex Zero Pair: Re="&X1$&", Im="&X2$ @ GOTO 1290
1560 !
1564 PR$=ZP$&" Pair #"&VAL$ (I)&", C0="&VAL$ (C0)&", C1="&VAL$ (C1)&", C2="
1566 PR$=PR$&VAL$ (C2)&", C3="&VAL$ (C3) @ GOSUB 1290
1568 !
1570 C(0,I)=C0 @ C(1,I)=C1 @ C(2,I)=C2 @ C(3,I)=C3 @ RETURN
1580 !
1590 ! ************** Butterworth Filter ****************************
1600 PR$=BF$ @ GOSUB 1720 ! Select High or Low Pass
1620 DISP "Enter";BW$;"(0-.5)";@ INPUT F3T
1630 IF F3T<= 0 OR F3T>= .5 THEN DISP "Use 0<bandwidth<.5" @ BEEP @ GOTO 1620
1640 OM=1/TAN (180*F3T) @ C0=1 @ C3=-(2*M)
1650 PR$=BF$&PR$&" pass filter, "&BW$&"="&VAL$ (F3T) @ GOSUB 1290
1660 IF N=1 THEN 1700 ELSE K1=OM+1/OM @ K2=2*(OM-1/OM)
1670 FOR I=1 TO N2 @ GK=I @ IF FLAG1 THEN GK=I-.5
1680   CG=2*COS (180*GK/N) @ CK=CG+K1 @ C1=(CG-K1)/CK @ C2=K2/CK @ GOSUB 1450
1690 NEXT I @ IF FLAG1 THEN RETURN
1700 RP=(OM-1)/(OM+1) @ GOTO 1945
1710 !
1720 ! ********** Sub to select High or Low Pass ******************
1725 PR$=PR$&" filter" @ GOSUB 1290 @ REF=0
1730 DISP "Enter 'H' for High Pass, or 'L' for Low Pass ";CC$;" Filter";
1740 INPUT R$@ R$=UPC$ (R$) @ IF R$="H" THEN M=1 @ PR$="High" @ GOTO 1760
1750 IF R$#"L" THEN DISP "Use H or L" @ BEEP @ GOTO 1730 ELSE M=-1 @ PR$="Low"
1760 RETURN
1770 !
1780 ! ************** Chebyshev Filter *****************************
1790 PR$=CC$ @ GOSUB 1720 ! H or L Pass
1810 DISP "Enter";BW$;" (0-.5) and p-p ripple (dB)";@ INPUT F3T,R
1820 IF F3T<= 0 OR F3T>= .5 THEN DISP "Use 0<bandwidth<.5" @ BEEP @ GOTO 1810
1830 R=ABS (R) @ R=MAX (EPS ,R) @ e2=10^(R/10)-1 @ e=SQR (e2) @ BE=1/e
1840 IF R<3 THEN C=FNCOSH(FNICOSH(BE)/N) ELSE C=COS (ACS (BE)/N)
1850 OM=C*TAN (180*F3T)^M @ C0=1 @ C3=-(2*M)
1860 PR$=CC$&" "&PR$&" Pass Filter, "&BW$&" = "&VAL$ (F3T)
1870 PR$=PR$&", Ripple="&VAL$ (R)&"dB, OM="&VAL$ (OM) @ GOSUB 1290
1880 AL=SQR (1+BE*BE) @ NI=1/N @ K1=(AL+BE)^NI @ K2=(AL-BE)^NI
1890 SH=(K1-K2)/2 @ IF N=1 THEN 1940 ELSE CH=SH+K2
1900 FOR I=1 TO N2 @ GAM=180*(I-.5)/N @ A=-(SIN (GAM)*SH) @ B=COS (GAM)*CH
1910   D=(A+OM)/B @ E=(A-OM)/B @ PHI=ATN (1/D)-ATN (1/E)
1920   C1=-((D*D+1)/(E*E+1)) @ C2=-(2*M*COS (PHI)*SQR (-C1)) @ GOSUB 1450
1930 NEXT I @ IF FLAG1 THEN RETURN
1940 RP=M*(SH-OM)/(SH+OM)
1945 CZ=-M @ RZ=M @ GOTO 1250
1950 !
1960 ! ************** Frequency Response Sweep *****************
1970 PR$="Frequency response--digital filter" @ GOSUB 1290
1975 IF TF>1 THEN 2020
```

```
1980 DISP "Enter norm ref freq (0-.5)";@ INPUT FT
1990 IF FT<0 OR FT>.5 THEN DISP "Use freq 0-.5" @ BEEP @ GOTO 1980
2000 GOSUB 2200 @ REF=FNPDB(GA)
2010 PR$="Reference gain="&VAL$ (FNRD(REF))&"dB, at fT="&VAL$ (FT) @ GOSUB 1290
2020 DISP "Enter min normalized freq, increment and max norm freq (0-.5)";
2030 INPUT XL,DX,XR@ NP=(XR-XL)/DX+1
2040 IF DX<= 0 THEN DISP "Use increment > 0 " @ BEEP @ GOTO 2020
2050 IF XL>XR THEN DISP "Use max>min" @ BEEP @ GOTO 2020
2055 IF NP>201 THEN DISP "Use # pts<202" @ BEEP @ GOTO 2020
2060 K=1 @ FOR FT=XL TO XR STEP DX
2070    GOSUB 2180 @ FF(K),AMDB=FNPDB(GA)-REF @ PR$="fT="&VAL$ (FT)&", Gain="
2080    PR$=PR$&VAL$ (FNRD(AMDB))&"dB"
2085    IF TF=1 THEN PR$=PR$&", Phase="&VAL$ (FNRD(PH))&"deg"
2090    GOSUB 1290 @ K=K+1
2095 NEXT FT
2100 DISP "Select Plot Mode:" @ DISP @ DISP " 1 No Plot" @ DISP " 2 New Plot"
2102 IF PF>1 THEN DISP " 3 Overlay Plot"
2104 DISP "Enter #";@ INPUT PF
2120 IF PF<1 OR PF>3 THEN DISP "Use # 1-3" @ BEEP @ GOTO 2104
2130 YB=-80 @ YT=10 @ S$=VAL$ (N)&" Delay-Line Filter"
2140 X$="Normalized frequency--fT" @ YL$="Relative Response (dB)"
2150 ON PF GOTO 1760,3190,3205
2160 !
2170 ! ************** sub to calc gain of filter  ************************
2180 ON TF GOTO 2200,2300,2330
2190 !
2200 FF=360*FT @ AM=1 @ PH=0 @ SI=SIN (FF) @ CS=COS (FF) @ IF N=1 THEN 2260
2210 CS2=COS (2*FF) @ SI2=SIN (2*FF)
2220 FOR I=1 TO N2
2230    RN=CS2+C(3,I)*CS+C(0,I) @ IN=SI2+C(3,I)*SI ! I,Q, num.
2240    RD=CS2-C(2,I)*CS-C(1,I) @ ID=SI2-C(2,I)*SI @ GOSUB 2270 ! I,Q, denom.
2250 NEXT I @ IF FLAG1 THEN RETURN
2260 IN,ID=SI @ RN=CS+CZ @ RD=CS-RP
2270 DEN=MAX(SQR(RD*RD+ID*ID),1E-20)
2275 PHN=FNPATN2(IN,RN) @ AMN=SQR (RN*RN+IN*IN) @ AM=AM*AMN/DEN
2280 GA=AM*AM @ PH=PH+PHN-FNPATN2(ID,RD) @ RETURN
2290 !
2300 ! ********* Simplified Gain for Butterworth  *************
2310 GA=1/(1+1/(OM*FNPTAN(180*FT))^(2*M*N)) @ RETURN
2320 !
2330 ! ************** Simplified Gain for Chebyshev  ***********
2340 X=OM/FNPTAN(180*FT)^M @ GOSUB 580 ! Cheby poly
2350 GA=1/(1+e2*TN*TN) @ RETURN
2360 !
2370 ! ************* Step or Impulse Response  ******************
2380 IFLAG=R-3 @ IF IFLAG THEN PR$="Step" ELSE PR$="Impulse"
2390 PR$=PR$&" Response" @ GOSUB 1290
2400 ESQ,TODD=0 @ E=1 @ IF N=1 THEN 2420
2410 FOR I=1 TO N2 @ T(1,I),T(2,I)=0 @ NEXT I ! Zero double delay lines
2420 FOR J=1 TO 5 @ IF N=1 THEN 2470
2430    FOR I=1 TO N2
2440       T2=T(1,I) @ T1=E+C(1,I)*T(2,I)+C(2,I)*T2 ! Next IPP's outputs
2450       E=T1+T2*C(3,I)+T(2,I)*C(0,I) @ T(1,I)=T1 @ T(2,I)=T2 ! output,advance
```

```
2460    NEXT I @ IF FLAG1 THEN 2480
2470    T=E+TODD*RP @ E=CZ*TODD+T @ TODD=T ! single delay line
2480    ESQ=ESQ+E*E @ PR$="Output = "&VAL$ (FNRD(FNVDB(E)))&"dB"
2490    IF IFLAG=0 THEN PR$=PR$&",  Noise Gain="&VAL$ (FNRD(FNPDB(ESQ)))&"dB"
2500    GOSUB 1290 @ E=IFLAG ! Input for next IPP
2510 NEXT J
2520 DISP "More responses";@ INPUT R$@ R$=UPC$ (R$) @ IF R$="N" THEN RETURN
2540 IF R$#"Y" THEN DISP "Y or N?" @ BEEP @ GOTO 2520 ELSE 2420
2550 !
2560 ! ***********************   Clutter Attn, I, Avg gain   *************
2570 IF TF=1 THEN FT=.5 @ GOSUB 2200 @ GG=GA @ GOTO 2615
2580 GG=1 @ IF M=1 THEN 2615
2590 DISP "Low pass ";CC$;" or ";BF$;"filter, do you wish to use same ";BW$;
2600 DISP " but High pass";@ INPUT R$@ R$=UPC$ (R$) @ IF R$="N" THEN 660
2610 IF R$#"Y" THEN DISP "Y or N?" @ BEEP @ GOTO 2590
2615 M=1 @ DISP "Calc Avg Gain"
2620 PSUM=0 @ FOR J=1 TO 25 ! Avg Gain
2630    FT=J/50 @ GOSUB 2170 @ PSUM=PSUM+GA/GG
2640 NEXT J @ GAVG=PSUM/25 @ PR$="Gavg="&VAL$ (FNRD(GAVG)) @ GOSUB 1280
2650 DISP "Select next operation" @ DISP
2660 DISP "1  Single";CB$;", calc CA, I"
2670 DISP "2  Sweep same"
2680 DISP "3  Main Menu"
2690 DISP @ DISP "Enter #";@ INPUT R@ IF R=3 THEN RETURN
2700 IF R=2 THEN 2810
2710 IF R#1 THEN DISP "Use # 1-3" @ BEEP @ GOTO 2690
2750 DISP "Enter normalized";CB$;" (0-.5)";@ INPUT DF@ GOSUB 2850 ! CA, I
2770 DISP "More Bandwidths";@ INPUT R$@ R$=UPC$ (R$) @ IF R$="N" THEN 2650
2790 IF R$#"Y" THEN DISP "Y or N?" @ BEEP @ GOTO 2770 ELSE 2750
2800 !
2810 DISP "Enter";CB$;": min, max, increment";@ INPUT DF1,DF2,DDF
2820 IF DDF<= 0 OR DF2<DF1 THEN DISP "Bad input" @ BEEP @ GOTO 2810
2830 FOR DF=DF1 TO DF2 STEP DDF @ GOSUB 2850 @ NEXT DF @ GOTO 2650
2840 !
2850 ! *****  Sub to calc Clutter Attenuation (CA), and Imp Fact. (I)*****
2860 DFP=MAX (.0000000001,DF) ! Protect /0
2870 KC=-(DFP*DFP/4/LOG (2)) @ FM=SQR (-(KC*LOG (100000000)))
2880 NSUM,DSUM=0 @ FOR FT=FM/15 TO FM STEP FM/15
2890    PSD=EXP (FT*FT/KC) @ GOSUB 2170 @ NSUM=NSUM+PSD @ DSUM=DSUM+GA*PSD/GG
2900 NEXT FT @ CA=NSUM/DSUM @ CADB=FNPDB(CA) @ IMPDB=CADB+FNPDB(GAVG)
2910 PR$=CB$&"="&VAL$ (DFP)&", CA="&VAL$ (FNRD(CADB))
2920 PR$=PR$&"dB, Imp Fact ="&VAL$ (FNRD(IMPDB))&"dB" @ GOTO 1280
2930 !
2940 ! ***********   INITIALIZATION &  PRINTER SELECTION   ***********
2950 PAGESIZE 24 @ CLEAR @ DEG @ PRINTER IS 1 @ IOFLAG=0
2960 DISP TAB (10);T$ @ DISP
2970 DISP "Enter today's date (any comma-less format)";@ INPUT DA$
2980 DISP "Select Printer:" @ DISP
2990 DISP "1  CRT"
3000 DISP "2  HP-82905B--HP-IB"
3010 DISP "3  HP-82905B--Centronics"
3020 DISP "4  MX-100"
```

```
3030 DISP "Enter #";@ INPUT PN@ PF=0
3040 IF PN<1 OR PN>4 THEN DISP "Use # 1-4 " @ BEEP @ GOTO 3030
3045 IF PN#1 THEN DISP "Make sure Top-Of-Form is set properly, Press [END
LINE]";@ INPUT PR$
3050 ON PN GOTO 3150,3070,3080,3120
3060 !
3070 PRINTER IS 301 @ ZZZ=3 @ GOTO 3110
3080 PRINTER IS 701 @ ZZZ=7
3090 ON ERROR GOTO 3110 @ a=BTD ("1") ! I/O ROM Test
3100 CONVERT OUT ZZZ PAIRS ; ZZZ$@ IOFLAG=1
3110 OFF ERROR @ PRINT CHR$ (27)&"&l1L" @ GOTO 3140
3120 PRINTER IS 701 @ PRINT CHR$ (27);"N";CHR$ (8)
3140 PRINT TAB (15);T$ @ PRINT TAB (25);DA$
3150 DISP @ DISP "Answer questions with 'Y' or 'N', [END LINE]" @ DISP @ RETURN
3160 !
3170 ! ****************** PLOT DATA ********************************
3180 ! INPUTS : XL=MIN X, XR=MAX X, YT=MAX Y, YB=MIN Y
3190 LT=1 @ DEG @ AP=0 @ GOTO 3220
3205 AP=1 @ DISP "Enter LINE TYPE # 1-8 " @ INPUT LT
3207 IF LT<1 OR LT>8 THEN DISP "Use # 1-8" @ BEEP @ GOTO 3205
3220 IF PA=1 THEN LIMIT 0,125,0,75 @ X2=155 @ X3=85 @ GRAPH @ GOTO 3250
3230 IF IOFLAG THEN CONVERT OUT ZZZ ! Stop 0 conv for ext plotter
3240 LIMIT 12,252,18,192 @ X2=134 @ X3=73
3250 IF AP=0 THEN GOSUB 3400
3260 LOCATE 18,X2,10,84 @ SETGU @ IF AP=1 THEN 3290
3270 MOVE X3,89 @ CSIZE 5 @ LDIR 0 @ LORG 4 @ LABEL S$ @ CSIZE 4
3280 MOVE X2,5 @ LORG 9 @ LABEL DA$
3290 SCALE XL,XR,YB,YT
3300 MOVE XL,FF(1) @ I=1 @ LINE TYPE LT
3310 FOR X=XL+DX TO XR STEP DX @ I=I+1 @ DRAW X,FF(I) @ NEXT X
3320 IF PA=1 THEN 3340 ELSE PEN 0 @ MOVE XR,YT
3330 IF IOFLAG THEN CONVERT OUT ZZZ PAIRS ; ZZZ$@ RETURN ELSE RETURN
3340 ALPHA @ DISP "Dump to Printer";@ INPUT R$
3350 R$=UPC$ (R$) @ IF R$="N" THEN RETURN
3360 IF R$#"Y" THEN DISP "Y or N?" @ BEEP @ GOTO 3340
3370 IF PN=1 THEN GOSUB 2980 @ GOTO 3370 ! Select printer(not CRT)
3380 IF PN=4 THEN DUMP GRAPHICS 0,0,0,-1 @ RETURN ELSE DUMP GRAPHICS @ RETURN
3390 !
3400 IF PA=1 THEN LIMIT 0,125,0,75 @ X2=155 @ X3=85 @ GRAPH @ GCLEAR @ FRAME
ELSE LIMIT 12,252,18,192 @ X2=134 @ X3=73
3410 LOCATE 18,X2,10,84 @ FRAME
3420 MOVE 4,50 @ LORG 4 @ LDIR 90 @ CSIZE 5,.7,30 @ LABEL YL$
3430 MOVE X3,94 @ CSIZE 8 @ LDIR 0 @ LABEL T$
3440 MOVE X3,5 @ LORG 6 @ CSIZE 5 @ LABEL X$
3450 SCALE XL,XR,YB,YT @ CSIZE 4 @ FXD 2,0
3460 LGRID -.01,2,XL,YB,5,5 @ RETURN
3470 !
3480 ! ********** Sub to Select plotter, load binary GDUMP *********
3490 DISP "Enter Plotter Address (CRT=1)";@ INPUT PA
3500 IF PA<1 THEN DISP "Use # >=1" @ BEEP @ GOTO 3490
```

```
3510 ON ERROR GOTO 3530 @ GDUMP @ GOTO 3580 ! Error if GDUMP not resident,
3520 !   displays "Copyright Hewlett-Packard 203.09", if resident
3530 ON ERROR GOTO 3560 @ LOADBIN "GDUMP" @ GOTO 3510 ! several errors possible
3540 ! ERRN=67 indicates GDUMP not on disc
3550 ! ERRN=130     "        bad disc, no disc, door open, etc.
3560 IF ERRN =67 OR ERRN =130 THEN DISP "Insert disc with binary GDUMP, press
[CONT]" @ PAUSE @ GOTO 3530
3570 DISP "Unknown error,ERRN=";ERRN ;" try again" @ GOTO 3490
3580 OFF ERROR @ PLOTTER IS PA @ RETURN
```

Program DFT

```
10  ! Pgm DFT, rev 2/27/84, Written by W. SKILLMAN
20  ! Based on Pgms 5-4,5-5 and 5-6 in Radar Calc...
30  ! Rabiner & Gold FORTRAN FFT pgm (credited to Cooley, Lewis
40  !     & Welch) translated to BASIC, used for FFT, when # of
50  !     points and # of filters are equal and a power of 2 (new)
60  !     Combining used for Real data, as in Prog. 5-4
70  ! OPTION: Binary pgm, FFTg, speeds up FFT x5 (HP only)
80  !       (used automatically if on disc)
100 ! Goertzel algorithm used for DFT otherwise (as in Prog. 5-5)
110 !
120 ! Types of Input Data:(Data correction new)
130 !    Complex data:
140 !     In-phase, Quad components, (numeric) (new)
150 !     Amplitude(numeric), Phase (deg)
160 !     Amplitude(dB), Phase(deg) (new)
170 !   Real Data:
180 !     Amplitude;(numeric)
190 !   Weighting Options: (wt array DIM 1 to NIN)
200 !     Unweighted
210 !     Hamming
220 !     Dolph-Chebyshev (new, from Prog. 5-6)
230 !     User-defined (new)
240 ! Outputs numeric  (new), dB, for Polar, numeric I and Q (new)
250 ! dB ref options: unity, peak output, peak resp, noise (new)
260 ! FFT run time (exclusive of I/O)=0.194*N*LOG(N) sec(w/o FFTg)
270 ! I/O Arrays dimensioned 1 to NIN for complex, to NIN/2+1 for real
280 ! Inverse transform is normalized by division by NIN (new)
290 ! I/O, Wt arrays dimensioned for 64K RAM, increase for more RAM
300 ! Other OPTIONS: I/O ROM, Printer,Plotter
310 !
320 ! Important variables:(all angles in deg)
330 !    AI      = I array, I/O
340 !    AM      = Amplitude array, I/O
350 !    ANG2    = 360/NIN
360 !    AQ      = Q array, I/O
370 !    ARG     = angle, FFT
380 !    B:        0=complex data, 1= Real data
390 !    DFLAG   = Data type flag 1=I,Q,2=Amp, Phase, 3=Amp(dB), Phase
400 !              4= Real
410 !    EFLAG:    0=NIN a power of 2 and # filt out=# points, else =1
420 !    FFLAG   = 1, Binary FFTg in, else =0
430 !    GSL     = sidelobe level(dB), Dolph-Chebyshev weights.
440 !    INFLAG:  1=odd index Data entry , 0=even
450 !    IOFLAG  = 1 if I/O ROM in and HP-82905 printer, else =0
460 !    MX      =dB reference(numeric)
470 !    MXDB    = dB ref (dB)
480 !    N       =NIN for Complex data, =NIN/2 for Real data
490 !    NIN     = # of points in DFT/FFT
500 !    NINV:    -1= Direct Transform, 1= Inverse
510 !    NL      =# of calls to Goertzel filter pair algorithm
520 !    NOUT    =# of filters: 0<=fT<=1
```

```
530 !    OFLAG:    0= # of DFT points even, 1= odd
540 !    PH       = Phase Array, I/O
550 !    RF       =dB ref flag: 1=unity, 2=Peak output, 3=sum of
560 !             abs(volts),4=SQR(sum of power)
575 !    SFFLAG   0=filt bank, 1=sing. filt.
580 !    SFLAG:    0=even no. of filters, 1=odd
590 !    SUM2     =power sum of input data
600 !    VSUM     =sum of abs of ampls of input voltage
610 !    WT       = weight array (1 to NIN)
620 !    ZFLAG:    0=no zero fill, 1= zero fill data array to end
630 !    ZZZ      =HP-82905B Printer Interface Code
640 !    ZZZ$     = normal to slashed zero conversion string
650 !
670 DIM AI(513),AQ(513),WT(1024),PH(513),AM(513),PR$[80],E$[70],HD$[70]
680 DIM A$[60],T$[40],DC$[30],FPT$[30],H$[30],S$[30],DD$[60],HI$[70]
690 !
700 INTEGER N,I,J,K,L,LE,LE1,NIN,ZFLAG,DFLAG,NINV,B,N1,OFLAG,NMAX
710 INTEGER INFLAG,RF,PN,PA,EFLAG,ZZZ
720 !
730 EE$="     fT   " @ ZZZ$="0"&CHR$ (210)
740 E$="        Ampl           Amp(dB)     Phase(deg)       I              Q"
750 HI$="   Sample"&E$ @ E$=EE$&E$
760 T$="***DISCRETE FOURIER TRANSFORM***"
770 DC$="Sidelobe level in dB" @ HA$="Hamming"
780 X$="Frequency*IPP=fT" @ DO$="Dolph-Chebyshev"
790 Y$="Amplitude(dB)" @ UN$="Unweighted"
800 UD$="User-defined" @ PT$="number of points"
810 FPT$="DFT, "&PT$ @ DA$=""
820 WT$=" Weighting" @ REF$="dB reference= "
830 AM$,AMP$="" @ IQMIN=SQR (EPS )
835 AI(1),AI(2),AQ(1),AQ(2)=0
840 !
850 ! *************    Defined Functions    *****************
860 DEF FNVDB(X) = 2*FNPDB(X) ! Voltage ratio to dB
870 DEF FNPDB(X) = 10*LGT (MAX (EPS ,ABS (X))) ! power ratio to dB
880 !
890 GOSUB 4580 @ GOSUB 4840 ! Init, printer& plotter selection,load GDUMP
900 !
910 T$="** SPECTRUM **" @ GOSUB 2690 ! Select type of data entry
920 IF DFLAG>3 THEN NMAX=1024 ELSE NMAX=512 ! RAM Dependent
930 DISP "Enter ";PT$;"(<=";NMAX;") ";@ INPUT NIN
940 IF NIN>NMAX OR NIN<2 THEN DISP "USE 2<=N<=";NMAX @ BEEP @ GOTO 930
950 M=LOG (NIN)/LOG (2) @ EFLAG=0
960 IF ABS (M-INT (M+.5))>.0001 THEN EFLAG=1 ! Set flg if NIN # Pwr of 2
970 OFLAG=2*FP (NIN/2) ! 0, Even; 1,Odd
980 ANG2=360/NIN @ N1=NIN-1 @ N=NIN
990 IF DFLAG>3 THEN N=NIN/2 @ M=M-1
1000 GOSUB 2870 @ GP=SUMWTS^2/SUMWTS2 @ LW=GP/NIN ! Select Wt. fn.
1010 SUMWTS2=SUMWTS2/NIN ! normalize weighting fn.
1020 PR$="Proc. gain = "&VAL$ (FNPDB(GP))&"dB,  Wt. loss="&VAL$ (FNPDB(LW))&"dB"
```

```
1030 GOSUB 4260 ! print Proc. Gain & Wt. Loss
1040 DISP @ IF DFLAG<4 THEN DISP "Enter complex data in pairs,separated by
comma ";ELSE DISP "Enter Real data singly, ";
1050 DISP "then press [END LINE]" @ DISP @ DISP "Enter Z";
1060 IF DFLAG<4 THEN DISP ", 0";
1070 DISP " to zero-fill to end and get next menu." @ DISP
1080 DISP " Data may be corrected after all entries are made, using next menu"
1090 IF PN#1 THEN PRINT @ PRINT "   Index";TAB (20);A$
1100 ZFLAG,VSUM,SUM2=0
1110 FOR I=1 TO N ! Data Entry
1120    IF ZFLAG THEN AI(I),AQ(I)=0 @ GOTO 1140
1130    ON DFLAG GOSUB 2080,1980,1980,2150 @ GOTO 1150
1140    IF DFLAG<4 THEN AM(I),PH(I)=0 ELSE AM(II+1),AM(II+2)=0 @ II=II+2
1150 NEXT I
1160 PR$="" @ GOSUB 4270 @ DISP "Correct Data or Start DFT " @ DISP
1170 DISP "1   Correct Data"
1180 DISP "2   Start DFT-Filter Bank"
1190 DISP "3   Start Inverse DFT-Filter Bank"
1200 DISP "4   Single Filter DFT"
1210 DISP "5   Main Menu"
1220 DISP @ DISP "Enter #";@ INPUT R
1230 IF R<1 OR R>5 THEN DISP "use # 1-5" @ GOTO 1220
1240 ON R GOTO 1270,1330,1340,3890,1870
1250 !
1260 ! *************** DATA CORRECTION ***************************
1270 DISP "Enter index (1-";NIN;") of data to correct([EL] to Quit)";@ INPUT I$
1280 IF I$="" THEN 1160 ELSE I=VAL (I$)
1290 IF I>NIN OR I<1 THEN DISP "USE 1<Index<=";NIN @ BEEP @ GOTO 1270
1300 ON DFLAG GOSUB 4340,4310,4310,4400 @ GOTO 1270 ! Data entry
1310 !
1320 ! ************* START DFT/FFT FILTER BANK ******************
1330 F$="Direct " @ HD$=E$ @ NINV=-1 @ H$=" filters" @ GOTO 1350
1340 F$="Inverse " @ HD$=HI$ @ NINV=1 @ H$=" time slots "
1350 DISP "Select Reference for dB" @ DISP
1360 DISP "1   Unity"
1370 DISP "2   Peak Output"
1380 DISP "3   Peak Response"
1390 DISP "4   Noise"
1400 DISP @ DISP "Enter #";@ INPUT RF@ YT=0
1410 IF RF<1 OR RF>4 THEN DISP "Use # 1-4" @ BEEP @ GOTO 1400
1420 ON RF GOTO 1440,1460,1480,1510
1430 !
1440 MX=SQR (SUMWTS2) @ G$="unity" @ YT=20*INT (LGT (VSUM))+10 @ GOTO 1530
1450 !
1460 G$="peak output" @ GOTO 1530
1470 !
1480 G$="peak response" @ MX=VSUM*SQR (SUMWTS2) @ IF NINV=1 THEN MX=MX/NIN
1490 GOTO 1530
1500 !
1510 G$="power sum(noise)" @ MX=SQR (SUM2*SUMWTS2) @ IF NINV=1 THEN MX=MX/NIN
1520 YT=10*INT (LGT (NIN))+10
1530 PR$=F$&FPT$&"="&VAL$ (NIN) @ GOSUB 4230 @ YB=YT-90
1540 PR$=REF$&G$ @ GOSUB 4230
```

```
1550 DISP "Select # of";H$;" to sweep, w/o foldover if Real"
1560 DISP TAB (5);"No entry defaults to ";NIN;H$;@ INPUT NOUT$
1570 NOUT=NIN @ IF NOUT$#"" THEN NOUT=VAL (NOUT$)
1580 IF NOUT#NIN THEN EFLAG=1
1590 START=TIME  @ ON EFLAG+1 GOSUB 2270,4000 ! FFT or DFT
1600 X=TIME -START @ PR$="DFT/FFT Processing time="&VAL$ (X)&"seconds"
1610 GOSUB 4230 @ DISP "Rect. to Polar";
1620 IF RF=2 THEN DISP " and finding max amp" @ MX=EPS
1630 NMAX=N+B @ IF EFLAG THEN NMAX=NOUT
1640 IF DFLAG>3 AND EFLAG THEN NMAX=NL
1645 DF=1/NMAX @ IF DFLAG>3 THEN DF=.5/(NMAX-1)
1650 FOR I=1 TO NMAX
1660    GOSUB 4520 @ IF RF=2 THEN MX=MAX (MX,AM(I)) ! R to P,find max amp.
1670 NEXT I @ MXDB=FNVDB(MX)
1680 PR$="dB output is normalized to "&G$ @ GOSUB 4260
1690 FOR2: IMAGE 3X,"Max Amplitude=",M4D.4D,"dB or =",MD.4DE,"(numeric)"
1700 DISP USING FOR2 ; MXDB,MX @ DISP @ DISP HD$
1710 IF PN#1 THEN PRINT USING FOR2 ; MXDB,MX @ PRINT @ PRINT HD$
1720 FOR1: IMAGE 2X,D.4D,3X,MD.4DE,2X,M5D.3D,2X,M4D.4D,2(2X,MD.4DE)
1730 FOR I=1 TO NMAX @ FT=I @ IF NINV=-1 THEN FT=(I-1)/NOUT
1740    AMDB=FNVDB(AM(I))-MXDB @ GOSUB 4470
1750 NEXT I
1760 FOR3: IMAGE /70("*")
1770 DISP USING FOR3 @ DISP @ IF PN#1 THEN PRINT USING FOR3 @ PRINT
1780 DISP @ DISP "Select Plot Option" @ DISP
1790 DISP "1  No Plot"
1800 DISP "2  Amplitude Plot"
1810 DISP "3  Amp + Phase Plot"
1820 DISP "Enter # ";@ INPUT R
1830 IF R<1 OR R>3 THEN DISP "Use # 1-3" @ BEEP @ GOTO 1820
1840 IF R>1 THEN GOSUB 3460
1850 !
1860 ! *********************** MAIN MENU  *******************
1870 DISP @ DISP "** Main Menu ** Select next operation" @ DISP
1880 DISP "1  New Data" @ DISP "2  New Data Type"
1890 DISP "3  New Weights" @ DISP "4  New # of Points"
1900 DISP "5  Start DFT" @ DISP "6  Start Inverse DFT" @ DISP "7  Disc Menu"
1910 DISP @ DISP "Enter # ";@ INPUT R@ IF R=7 THEN CHAIN "Autost"
1920 IF R<1 OR R>7 THEN DISP "USE # 1-7 " @ BEEP @ GOTO 1910
1930 ON R GOTO 1040,910,1000,920,1330,1340
1940 !
1950 ! **************   DATA ENTRY SUBROUTINES   ***********
1960 !
1970 ! *************  AMPLITUDE, PHASE DATA ENTRY  *********
1980 DISP "Enter Amp(";C$;"), Phase(deg), i=";I;@ INPUT AMP$,PH
1990 IF UPC$ (AMP$)="Z" THEN ZFLAG=1 @ AI(I),AQ(I),AM(I),PH(I)=0 @ RETURN
2000 ON ERROR GOTO 1980 @ AM=VAL (AMP$)
2010 OFF ERROR @ AN=AM @ IF DFLAG=3 THEN AN=10^(AM/20)
2020 X=AN*WT(I) @ AI(I),AM(I)=X*COS (PH)
2030 AQ(I),PH(I)=X*SIN (PH)
2040 IF PN#1 THEN PRINT TAB (3);I,AM,PH
2050 VSUM=VSUM+ABS (AN) @ SUM2=SUM2+AN*AN @ RETURN
2060 !
```

```
2070 ! ***************  I, Q DATA ENTRY  **********************
2080 DISP "Enter I(REAL), Q(IMAG),i=";I;@ INPUT AI$,AQ
2090 IF UPC$ (AI$)="Z" THEN ZFLAG=1 @ AI(I),AQ(I),AM(I),PH(I)=0 @ RETURN
2100 ON ERROR GOTO 2080 @ AI=VAL (AI$) @ IF PN#1 THEN PRINT TAB (3);I,AI,AQ
2110 OFF ERROR @ SS=AI*AI+AQ*AQ @ SUM2=SUM2+SS @ VSUM=VSUM+SQR (SS)
2120 AI(I),AM(I)=AI*WT(I) @ AQ(I),PH(I)=AQ*WT(I) @ RETURN
2130 !
2140 ! ***********  REAL DATA ENTRY  *************************
2150 INFLAG=1 @ AI(I),AQ(I)=0 @ II=2*I-2
2160 II=II+1
2170 IF II<= NIN THEN DISP "Enter Amp(";C$;"),i=";II;@ INPUT AM$ ELSE RETURN
2180 IF UPC$ (AM$)#"Z" THEN 2220
2190 ZFLAG=1 @ AM(II)=0 @ IF INFLAG THEN II=II+1 @ AM(II)=0
2200 RETURN
2210 !
2220 ON ERROR GOTO 2170 @ AN=VAL (AM$) @ IF PN#1 THEN PRINT TAB (3);II,AN
2230 OFF ERROR @ AM(II),X=AN*WT(II) @ GOSUB 2050
2240 IF INFLAG THEN AI(I)=X @ INFLAG=0 @ GOTO 2160 ELSE AQ(I)=X @ RETURN
2250 !
2260 ! *******   Calculate FFT using  FFTg or Rabiner& Gold  *******
2270 IF FFLAG=0 THEN 2340
2280 DISP "Performing FFTg" @ RAD @ FFT AI,AQ,N,M,-NINV @ DEG @ GOTO 2540
2330 !
2340 DISP "Re-ordering data" @ J,LE=1 @ DF=1/NIN
2350 FOR I=1 TO N-1
2360   IF I>= J THEN 2390
2370   TI=AI(J) @ AI(J)=AI(I) @ AI(I)=TI ! Swap AI(J),AI(I)
2380   TQ=AQ(J) @ AQ(J)=AQ(I) @ AQ(I)=TQ ! Swap AQ(J),AQ(I)
2390   K=N/2
2400   IF K<J THEN J=J-K @ K=K/2 @ GOTO 2400 ELSE J=J+K
2410 NEXT I @ DISP @ DISP "Performing FFT" @ ARG=NINV*360
2420 FOR L=1 TO M ! Start of FFT using Rabiner & Gold pgm
2430   LE1=LE @ LE=LE*2 @ UI=1 @ UQ=0 @ ARG=ARG/2
2440   WI=COS (ARG) @ WQ=SIN (ARG)
2450   FOR J=1 TO LE1
2460     FOR I=J TO N STEP LE @ IPP=I+LE1
2470       TI=AI(IPP)*UI-AQ(IPP)*UQ
2480       TQ=AI(IPP)*UQ+AQ(IPP)*UI
2490       AI(IPP)=AI(I)-TI @ AQ(IPP)=AQ(I)-TQ
2500       AI(I)=AI(I)+TI @ AQ(I)=AQ(I)+TQ
2510     NEXT I @ T=UI @ UI=T*WI-UQ*WQ @ UQ=T*WQ+UQ*WI
2520   NEXT J @ DISP " Finished loop # ";L
2530 NEXT L
2540 IF DFLAG<4 THEN RETURN
2550 !
```

```
2560 ! *********  COMBINE REAL DATA  *********
2570 DISP @ DISP "Combining real data"
2580 AI(N+1)=AI(1)-AQ(1) @ AI(1)=AI(1)+AQ(1) @ AQ(1),AQ(N+1),ANGR=0
2590 FOR I=2 TO N/2
2600   ANGR=ANGR-ANG2 @ J=N-I+2
2610   Y=AI(J)+AI(I) @ X=AI(J)-AI(I)
2620   T=AQ(J)+AQ(I) @ U=AQ(I)-AQ(J) @ PH=0
2630   AMP=SQR (X*X+T*T)/2 @ IF AMP THEN PH=ATN2 (X,T)+ANGR
2640   AQ(I)=U/2+AMP*SIN (PH) @ AQ(J)=AQ(I)-U
2650   AI(I)=Y/2+AMP*COS (PH) @ AI(J)=Y-AI(I)
2660 NEXT I @ RETURN
2670 !
2680 ! *****   SELECT TYPE OF DATA, SET DFLAG, AND C$   *******
2690 DISP @ DISP "Select type of data entry" @ DISP @ B=0
2700 DISP "1  I, Q (numeric)"
2710 DISP "2  Amp(num), Phase (deg)"
2720 DISP "3  Amp(dB), Phase (deg)"
2730 DISP "4  Real(num)"
2740 DISP @ DISP "Enter #";@ INPUT R
2750 IF R<1 OR R>4 THEN DISP "Use # 1-4 " @ BEEP @ GOTO 2740
2760 DFLAG=R @ ON R GOTO 2780,2800,2820,2840
2770 !
2780 A$="      REAL(I)              IMAG(Q)" @ RETURN
2790 !
2800 A$="   Amplitude               Phase(Deg)" @ C$="num" @ RETURN
2810 !
2820 A$="   Ampl(dB)                Phase(Deg)" @ C$="dB" @ RETURN
2830 !
2840 A$="   Amplitude" @ C$="Enter Amp(num),i=" @ B=1 @ RETURN
2850 !
2860 ! *********  SELECT WEIGHTING  ************************
2870 DISP @ DISP "Select ";WT$ @ DISP @ SUMWTS,SUMWTS2=0
2880 DISP "1  ";UN$ @ DISP "2  ";HA$
2890 DISP "3  ";DO$ @ DISP "4  ";UD$
2900 DISP @ DISP "Enter # ";@ INPUT R
2910 IF R<1 OR R>4 THEN DISP "Use # 1-4" @ BEEP @ GOTO 2900
2920 ON R GOTO 2950,2990,3190,3060
2930 !
2940 ! ***************  UNWEIGHTED  ********************
2950 FOR I=1 TO NIN @ WT(I)=1 @ NEXT I @ SUMWTS,SUMWTS2=NIN
2960 DD$,PR$="    "&UN$ @ GOTO 4270
2970 !
2980 ! ***************  HAMMING WEIGHTS  *****************
2990 DD$,PR$="    "&HA$ @ GOSUB 4270 @ ANG=ANG2/2
3000 FOR J=1 TO NIN/2+.5 @ X=1-.8575*COS (ANG)
3010   GOSUB 3350 @ J=J-1 @ ANG=ANG+ANG2
3020 NEXT J @ SUMWTS2=SUMWTS2*2-OFLAG*X*X
3030 SUMWTS=SUMWTS*2-OFLAG*X @ RETURN
3040 !
3050 ! ********  USER-DEFINED WEIGHTS  ***********
3060 DD$,PR$="    "&UD$ @ GOSUB 4230
3070 H$="  Index              weight" @ GOSUB 4270
```

```
3080 FOR I=1 TO NIN @ GOSUB 3150 @ NEXT I
3090 DISP "Correction";@ INPUT R$@ R$=UPC$ (R$) @ IF R$="N" THEN RETURN
3100 IF R$#"Y" THEN DISP "Y or N?" @ BEEP @ GOTO 3090
3110 DISP "Enter Index, 1-";NIN;@ INPUT I
3120 IF I<1 OR I>NIN THEN DISP "Use 1<=Index<=";NIN @ BEEP @ GOTO 3090
3130 SUMWTS2=SUMWTS2-WT(I)^2 @ SUMWTS=SUMWTS-ABS (WT(I)) @ GOSUB 3150
3135 GOTO 3090
3140 !
3150 DISP "Enter WT(";I;")";@ INPUT X@ SUMWTS2=SUMWTS2+X*X @ WT(I)=X
3160 SUMWTS=SUMWTS+ABS (X) @ IF PN#1 THEN PRINT TAB (3);I;X
3165 RETURN
3170 !
3180 ! ****************** DOLPH-CHEBYSHEV WEIGHTS  ****************
3190 DISP "Enter ";DC$;
3200 INPUT X@ GSL=ABS (X)
3210 PR$,DD$=DO$&", "&DC$&"=-"&VAL$ (GSL) @ GOSUB 4270
3220 R=10^(GSL/20) @ R1=R+SQR (R*R-1) @ X=R1^(2/N1)
3230 X=(X-1)/(X+1) @ R4=X*X @ XN=R4*N1
3240 KK,K=INT (NIN/2+.5) @ J=1 @ GOSUB 3380
3250 XN=XN/X @ X=1/X @ GOSUB 3340
3260 IF K=2 THEN 3310 ELSE X=XN @ GOSUB 3340
3270 IF K=3 THEN 3310
3280 FOR K=3 TO KK-1
3290   GOSUB 3380 @ GOSUB 3340
3300 NEXT K
3310 X=1 @ GOSUB 3340 @ SUMWTS2=SUMWTS2*2-OFLAG
3320 SUMWTS=SUMWTS*2-OFLAG @ RETURN
3330 !
3340 ! SUB to store weights and sum power
3350 WT(J),WT(NIN-J+1)=X @ SUMWTS2=SUMWTS2+X*X @ J=J+1
3360 SUMWTS=SUMWTS+X @ RETURN
3370 !
3380 R10,X=1 ! SUB To calc  one coef.
3390 FOR I=1 TO K-2
3400   X=X*R4*(K-1-I)*(NIN-K-I)/(I+I*I) @ R10=R10+X
3410   IF ABS (X)/R10<= 1.E-13 THEN 3430
3420 NEXT I
3430 X=R10*XN @ RETURN
3440 !
3450 ! ****************** SUB TO PLOT SPECTRUM  ********************
3460 IF PA=1 THEN LIMIT 0,125,0,75 @ X2=145 @ X3=80 @ GOTO 3490 ! CRT Plot
3470 IF IOFLAG THEN CONVERT OUT ZZZ ! Stop 0 conversion for ext plotter
3480 LIMIT 0,240,0,190 @ X2=110 @ X3=62
3490 DISP "Enter subtitle(<31 CHAR.)";@ INPUT S$
3500 X4=1 @ IF DFLAG>3 THEN X4=.5
3510 GCLEAR @ DEG @ LOCATE 15,X2,10,84 @ FXD 1/X4,0
3520 MOVE 4,50 @ LORG 4 @ LDIR 90 @ CSIZE 5,.7,20 @ LABEL Y$
3530 NPTS$=VAL$ (NIN)&" Point DFT"
3540 MOVE 15,87 @ CSIZE 4 @ LDIR 0 @ LORG 3 @ LABEL NPTS$
3550 MOVE X2,87 @ LORG 9 @ LABEL REF$,G$
3560 MOVE X3,94 @ CSIZE 8 @ LORG 4 @ LABEL T$
3570 MOVE X3,89 @ CSIZE 5 @ LABEL S$
3580 MOVE X3,5 @ LORG 6 @ LABEL X$
3590 MOVE X2,5 @ CSIZE 4 @ LORG 9 @ LABEL DA$
```

```
3600 IF R=3 THEN MOVE 15,5 @ LORG 3 @ LABEL "---=Phase"
3610 SCALE 0,X4,YB,YT @ X5=1.13*X4
3620 LGRID -(X4/50),2,0,YB,5,5
3630 AXES X4/50,2,X4,0,5,5,4
3640 AMDB=FNVDB(AM(1))-MXDB
3650 MOVE 0,AMDB @ FT=0
3660 FOR I=2 TO NMAX
3670   FT=FT+DF @ DRAW FT,FNVDB(AM(I))-MXDB
3680 NEXT I
3690 IF DFLAG<4 THEN DRAW 1,AMDB
3700 IF R=2 THEN 3800
3710 SCALE 0,X4,-200,250 @ LORG 8
3720 FOR PH=250 TO -200 STEP -50
3730   MOVE X5,PH @ LABEL PH
3740 NEXT PH @ MOVE X5,0 @ LORG 6 @ LDIR 90 @ CSIZE 5,.7,20
3750 LABEL "Phase (deg)"
3760 MOVE 0,PH(1) @ FT=0 @ LINE TYPE 4
3770 FOR I=2 TO NMAX
3780   FT=FT+DF @ DRAW FT,PH(I)
3790 NEXT I @ IF DFLAG<4 THEN DRAW 1,PH(1)
3800 IF IOFLAG THEN CONVERT OUT ZZZ PAIRS ; ZZZ$
3810 IF PA#1 THEN PEN 0 @ MOVE 1,0 @ RETURN
3820 ALPHA @ DISP "Dump to Printer ";@ INPUT R$
3830 R$=UPC$ (R$) @ IF R$="N" THEN RETURN
3840 IF R$#"Y" THEN DISP "Y or N?" @ BEEP @ GOTO 3820
3850 IF PN=1 THEN GOSUB 4620 @ GOTO 3850 ! select printer
3860 IF PN=4 THEN DUMP GRAPHICS 0,0,0,-1 @ RETURN ELSE DUMP GRAPHICS @ RETURN
3870 !
3880 ! ***********  Single filter, using Goertzel Algorithm ********
3890 CLEAR @ NM=SQR (SUM2*SUMWTS2) @ PR$="dB norm to noise, using a factor of "
3900 PR$=PR$&VAL$ (NM)&"(numeric), or "&VAL$ (FNVDB(NM))&"(dB)" @ GOSUB 4260
3910 DISP "Enter normalized freq. fT,(1 to QUIT)";@ INPUT FT
3920 IF FT=1 THEN 1870
3930 IF ABS (FT)>1 THEN DISP "Use 0<=fT<=1" @ BEEP @ GOTO 3910
3940 SFFLAG=1 @ J=0 @ NINV=-1 @ GOSUB 4040 @ I=0 @ GOSUB 4520
3950 AMDB=-499 @ IF ABS (AM(0))>NM*EPS  THEN AMDB=FNVDB(AM(0)/NM)
3960 PR$="FREQ="&VAL$ (FT)&"  RESPONSE="&VAL$ (AMDB)&" dB,  PHASE="
3970 PR$=PR$&VAL$ (PH(0))&" DEG." @ GOSUB 4260 @ GOTO 3910
3980 !
3990 ! ********   FILTER BANK (DFT) (Goertzel) ********
4000 DF=1/NOUT @ FT,SFFLAG=0
4010 SFLAG=2*FP (NOUT/2) @ NL=1+INT (NOUT/2)
4020 FOR J=1 TO NL @ GOSUB 4040 @ FT=FT+DF @ NEXT J @ RETURN
4030 !
4040 ! ******  Filter pair, Goertzel Algo.(DFT)  *******
4050 Q2,Q3,I2,I3=0 @ PHF=-(NINV*360*FT) @ FF=2*COS (PHF)
4060 FOR I=1 TO NIN
4070   I1=AM(I)-I3+FF*I2 @ I3=I2 @ I2=I1 @ IF DFLAG>3 THEN 4090
4080   Q1=PH(I)-Q3+FF*Q2 @ Q3=Q2 @ Q2=Q1
4090 NEXT I @ IF DFLAG>3 THEN 4190 ELSE AMP=SQR (Q3*Q3+I3*I3)
4100 PH=0 @ IF AMP THEN PH=ATN2 (Q3,I3)
4110 PH1=PH-PHF @ K=J @ GOSUB 4160
```

```
4120 IF FT=0 OR SFFLAG THEN RETURN !    single filter
4130 IF J=NL THEN RETURN ! single filter at fT=.5, NIN even
4140 PH1=PH+PHF @ K=NOUT-J+2 !          complex conj. pair otherwise
4150 !
4160 AI(K)=I1-AMP*COS (PH1) ! complex data
4170 AQ(K)=Q1-AMP*SIN (PH1) @ RETURN
4180 !
4190 AI(J)=I1-I3*COS (PHF) ! real data
4200 AQ(J)=-(I3*SIN (PHF)) @ RETURN
4210 !
4220 ! ********  DISPLAY/PRINT/SPACE  *********
4230 DISP PR$ @ DISP @ IF PN#1 THEN PRINT PR$ @ PRINT
4240 RETURN
4250 !
4260 ! *****  DISPLAY/PRINT  ********
4270 DISP PR$ @ IF PN#1 THEN PRINT PR$
4280 RETURN
4290 !
4300 ! ********  CORRECT AMP, PHASE INPUT ***********
4310 DISP "Enter Amp(";C$;"),Phase(deg),i=";I;@ INPUT AM,PH
4320 GOSUB 4370 @ GOTO 2010
4330 !
4340 ! ******  CORRECT I,Q,ENTRY ***********
4350 DISP "Enter I(Real),Q(Imag),i=";I;@ INPUT AI$,AQ@ GOSUB 4370 @ GOTO 2100
4360 !
4370 A=(AI(I)^2+AQ(I)^2)/WT(I)^2 @ VSUM=VSUM-SQR (A) @ SUM2=SUM2-A @ RETURN
4380 !
4390 ! ******  CORRECT REAL ENTRY ********
4400 II=I @ Y=I/2+.5 @ I=INT (Y)
4410 DISP "Enter Amp(";C$;"),i=";II;@ INPUT AN@ A=AM(II)/WT(II)
4420 VSUM=VSUM-ABS (A)+ABS (AN) @ SUM2=SUM2-A*A+AN*AN @ X,AM(II)=AN*WT(II)
4430 IF FP (Y) THEN AQ(I)=X ELSE AI(I)=X
4440 RETURN
4450 !
4460 ! ******  PRINT/DISPLAY OUTPUT DATA ************
4470 DISP USING FOR1 ; FT,AM(I),AMDB,PH(I),AI(I),AQ(I)
4480 IF PN#1 THEN PRINT USING FOR1 ; FT,AM(I),AMDB,PH(I),AI(I),AQ(I)
4490 RETURN
4500 !
4510 ! ********  I,Q to Polar conv.  **********
4520 X=AI(I) @ Y=AQ(I) @ NNN=NIN @ IF NINV=-1 THEN NNN=1
4530 AAMIN=NNN*IQMIN @ IF ABS (X)>AAMIN THEN X=X/NNN ELSE X=IQMIN*SGN (X)
4540 IF ABS (Y)>AAMIN THEN Y=Y/NNN ELSE Y=IQMIN*SGN (Y)
4550 AM(I)=SQR (X*X+Y*Y) @ PH(I)=0 @ IF AM(I) THEN PH(I)=ATN2 (Y,X)
4560 AI(I)=X @ AQ(I)=Y @ RETURN
4570 !
4580 ! ********  INITIALIZATION &  PRINTER SELECTION  ******
4590 PAGESIZE 24 @ CLEAR @ DEG @ PRINTER IS 1 @ IOFLAG=0
4600 DISP TAB (10);T$ @ DISP
4610 DISP "Enter today's date, (any comma-less format)";@ INPUT DA$
4620 DISP "Select Printer:" @ DISP
4630 DISP "1   CRT"
4640 DISP "2   HP-82905B--HP-IB"
4650 DISP "3   HP-82905B--Centronics"
```

```
4660 DISP "4  MX-100"
4670 DISP "Enter #";@ INPUT PN
4680 IF PN<1 OR PN>4 THEN DISP "Use # 1-4 " @ BEEP @ GOTO 4670
4685 IF PN#1 THEN DISP "Make sure Top-Of-Form is set properly, Press [END
LINE]";@ INPUT PR$
4690 ON PN GOTO 4810,4710,4720,4780
4700 !
4710 PRINTER IS 301 @ ZZZ=3 @ GOTO 4730
4720 PRINTER IS 701 @ ZZZ=7
4730 ! For HP-82905B Printer, if I/O ROM, then convert normal zero to
4740 !    slashed zero.  (MX-100 can use DIP switch)
4750 ON ERROR GOTO 4770 @ a=BTD ("1") ! I/O ROM test
4760 CONVERT OUT ZZZ PAIRS ; ZZZ$@ IOFLAG=1 ! sets conv and flag if I/O ROM
4770 OFF ERROR @ PRINT CHR$ (27)&"&11L" @ GOTO 4800
4780 PRINTER IS 701 @ PRINT CHR$ (27);"N";CHR$ (8)
4800 PRINT TAB (15);T$ @ PRINT TAB (25);DA$
4810 DISP @ DISP "Answer questions with 'Y' or 'N', [END LINE]" @ DISP
4820 RETURN
4830 !
4840 ! ****** Sub to Select plotter, load binaries GDUMP,FFTg *****
4850 DISP "Enter Plotter Address (CRT=1)";@ INPUT PA
4860 IF PA<1 THEN DISP "Use # >=1" @ BEEP @ GOTO 4850
4870 ON ERROR GOTO 4880 @ GDUMP @ GOTO 4930 ! Error if GDUMP not IN
4880 ON ERROR GOTO 4910 @ LOADBIN "GDUMP" @ GOTO 4870 ! several errors possible
4890 ! ERRN=67 indicates GDUMP not on disc
4900 ! ERRN=130   "   bad disc, no disc, door open, etc.
4910 IF ERRN =67 OR ERRN =130 THEN DISP "Insert disc with binary GDUMP, press
[CONT]" @ PAUSE @ GOTO 4880
4920 DISP "Unknown error,ERRN=";ERRN ;" try again" @ GOTO 4850
4930 PLOTTER IS PA @ FFLAG=0
4932 ON ERROR GOTO 4940 @ FFT AI,AQ,2,1,1 @ GOTO 4950 ! Test for FFTg
4940 ON ERROR GOTO 4960 @ LOADBIN "FFTg"
4950 FFLAG=1 @ DISP "FFTg loaded"
4960 OFF ERROR @ RETURN
```

Program DOLPH

```
10 !  Pgm DOLPH, Rev 2/27/84, by W. Skillman
30 !     Derived from pgm 5-6 OF "RADAR CALC.
40 !
50 ! FUNCTIONS AVAILABLE:
60 !   SPECTRUM OF DOLPH-CHEBYSHEV WINDOWED DFT (PRINT/PLOT)
70 !   DOLPH-CHEBYSHEV DFT COEFS ,PROC GAIN, WT LOSS
80 !   CHEBYSHEV POLYNOMIAL; SINGLE VALUE OR SWEEP
90 !   COSH, INVERSE COSH
95 !
100 ! Improvements vs. Radar Calc.....
110 !    CRT or Plotter plot of Spectrum(CRT dump to printer also)
120 !
130 ! Important variables:
140 !    DFT.....increment of fT
150 !    FT .....freq/PRF
160 !    GGDB....response array (dB)
170 !    GSL.....sidelobe-to-mainlobe ratio (dB)
180 !    NPTS....# of points in DFT
190 !    PN......printer code
200 !    PLFLAG..plot flag
210 !    PRFLAG..print flag
220 !    r.......mainlobe-to-sidelobe ratio (numeric)
230 !    WT......weight array
240 !            WT(1)=WT(NPTS)=Edge Coefficient
250 !            WT(INT((NPTS+1)/2))=1=Center coef (odd) or pair(even)
260 !
270 DIM WT(1024),H$[60],B$[40],C$[60],T$[40],P$[35],Q$[35]
280 DIM X$[35],S$[50],GGDB(1024),PR$[60],DC$[35]
290 !
300 INTEGER NPTS,N1,K,KK,I,J,CFLAG,PFLAG,MCP,NPP,PN,NR
310 DATA No. of points in DFT,sidelobe level
320 DATA Processing gain=,Weighting loss=,**DOLPH-CHEBYSHEV WEIGHTING**
330 DATA Chebyshev Polynomial; order=,Frequency * IPP=fT,Response (dB)
340 DC$="Dolph-Chebyshev Coefficients"
350 H$="    fT              G(voltage)              G(dB)"
360 C$="K                   W(K)                    W(K)-dB"
380 Q$=" X                  T(X)"
400 READ B$,A$,D$,E$,T$,P$,X$,YL$
410 !
420 ! ***************  Defined Functions  *********************
430 DEF FNPDB(X) = 10*LGT (MAX (EPS ,ABS (X))) ! Power ratio to dB
440 DEF FNVDB(X) = 2*FNPDB(X) !              Voltage ratio to dB
450 DEF FNXZ(r,N) = FNCOSH(FNICOSH(r)/(N-1))
460 DEF FNCOSH(X) = (EXP (X)+EXP (-X))/2 !    Hyp Cosine
470 DEF FNICOSH(X) = LOG (ABS (X)+SQR (X*X-1)) ! Inv Hyp Cosine
480 !
490 GOSUB 2400 @ GOSUB 2630 @ T$="**DFT SPECTRUM**"
500 !
```

```
510 ! *******************   MAIN MENU  **************************
520 PR$="" @ GOSUB 2340 @ DISP "**MAIN MENU**: Select next operation" @ DISP
530 DISP "1 ";DC$;" or Response"
540 DISP "2 Cheby Poly Single Value"
550 DISP "3 Cheby Poly Sweep"
560 DISP "4 COSH(x)"
570 DISP "5 INV COSH(x)" @ DISP "6 Disc Menu"
580 DISP @ DISP "Enter #" @ INPUT NR@ IF NR=6 THEN CHAIN "Autost"
590 IF NR<1 OR NR>6 THEN DISP "Use # 1-6" @ GOTO 580
600 DISP @ ON NR GOTO 630,1680,1760,1910,1960
610 !
620 ! *********  MENU: CALC COEFS, OR FREQ RESPONSE **************
630 GOSUB 830 ! Enter # of points
640 GOSUB 890 ! Enter sidelobe level
650 PRFLAG=0 @ PLFLAG=1 @ DISP "Select next operation" @ DISP
660 DISP " 1 New ";B$
670 DISP " 2 New ";A$
680 DISP " 3 Calc. Coefs."
690 DISP " 4 Single freq. Response"
700 DISP "   Frequency response sweep"
710 DISP " 5    Display only"
720 IF PN=1 THEN 760
730 DISP "  6    Print"
740 DISP "  7    Print & Plot"
760 DISP "  8    Plot"
780 DISP " 9   Main Menu"
790 DISP @ DISP "Enter #";@ INPUT NR@ DISP
800 IF NR<1 OR NR>9 THEN DISP "Use # from 1-9" @ GOTO 790
810 ON NR GOTO 950,640,980,1350,1500,1460,1450,1490,520
820 !
830 ! *************  SUB TO ENTER # OF POINTS IN DFT  ***********
840 DISP "Enter ";B$;" (<=2047)";@ INPUT X@ N1=X-1
850 IF X<3 OR X>2047 THEN DISP "Use 3<=N<=2047" @ GOTO 840
860 IF FP (X) THEN DISP "Use integer for # of points" @ BEEP @ GOTO 840
870 IF PN#1 THEN PRINT B$;" = ";X
875 NPTS=X @ RETURN
880 !
890 ! *************  SUB TO ENTER SIDELOBE LEVEL  *************
900 DISP "Enter ";A$;" in dB";@ INPUT X@ GSL=-ABS (X)
910 r=10^(-(GSL/20)) @ R1=r+SQR (r*r-1)
920 PR$=A$&"= "&VAL$ (GSL)&"dB" @ GOTO 2300
930 !
940 ! *************   new # of points  ***************
950 GOSUB 830 @ GOTO 650
960 !
970 ! ******* SEGMENT TO CALC DOLPH-CHEBYSHEV WEIGHTS OR COEFS. ******
980 PR$=DC$ @ GOSUB 2300
990 X=R1^(2/N1)
1000 R4=((X-1)/(X+1))^2 @ XN=R4*N1
1010 X=NPTS/2 @ J=1
```

```
1020 R12,R11=-FP (X) ! Correct sums for odd # of points
1030 KK,K=INT (X+.5)
1040 GOSUB 1210 @ XN=XN/X
1050 PR$=C$ @ GOSUB 2330
1060 X=1/X @ GOSUB 1290 @ IF K=2 THEN 1130
1080 X=XN @ GOSUB 1290 @ IF K=3 THEN 1130
1100 FOR K=3 TO KK-1
1110    GOSUB 1210 @ GOSUB 1290
1120 NEXT K
1130 X=1 @ GOSUB 1290 @ X=R11*R11*2/R12
1160 PG=FNPDB(X) @ PR$=D$&VAL$ (X)&" or ="&VAL$ (PG)&"dB"
1170 PRINT @ DISP @ GOSUB 2340
1180 PR$=E$&VAL$ (PG-FNPDB(NPTS))&"dB" @ GOSUB 2340
1190 PR$="" @ GOSUB 2300 @ GOTO 650
1200 !
1210 ! **************    SUB TO CALC ONE COEF.    **************
1220 R10,X=1
1230 FOR I=1 TO K-2
1240    X=X*R4*(K-1-I)*(NPTS-K-I)/(I+I*I) @ R10=R10+X
1250    IF ABS (X)/R10<= 1.E-13 THEN 1270
1260 NEXT I
1270 X=R10*XN @ RETURN
1280 !
1290 ! **** SUB TO UPDATE SUMS, PRINT COEF AND STORE IN WT ARRAY *****
1300 R11=R11+X @ R12=R12+X*X
1310 WT(J),WT(NPTS-J+1)=X @ Y=FNVDB(X)
1320 DISP J,X,Y @ IF PN#1 THEN PRINT J,X,Y
1330 J=J+1 @ RETURN
1340 !
1350 ! ***** SUB TO CALC DFT RESPONSE FOR SINGLE FREQUENCY *******
1360 DISP "Enter normalized freq, fT";@ INPUT FT@ GOSUB 1640
1370 PR$="    fT="&VAL$ (FT)&"    G="&VAL$ (G)&"    G(dB)="&VAL$ (GDB)
1380 GOSUB 2340
1390 DISP "More freqs";@ INPUT R$@ R$=UPC$ (R$) @ IF R$="Y" THEN 1360
1410 IF R$="N" THEN 650 ELSE DISP "Y or N?" @ BEEP @ GOTO 1390
1430 !
1440 ! ****************    DFT Response Sweep    **************
1450 PLFLAG=2
1460 PRFLAG=1 @ GOTO 1500
1490 PLFLAG=2
1500 DISP @ DISP "Enter freq. increment,";CHR$ (7);"fT"
1510 DISP "    (.002 min for CRT Plot, .001 for Plotter)";@ INPUT DFT
1520 IF DFT<= 0 THEN DISP "Use # > 0" @ BEEP @ GOTO 1500
1530 PR$=T$ @ GOSUB 2370
1540 DISP @ IF PRFLAG=1 AND PN#1 THEN PRINT
1550 PR$=H$ @ GOSUB 2370
1560 I=1 @ FOR FT=0 TO .5 STEP DFT
1570    GOSUB 1640 @ GGDB(I)=GDB @ I=I+1
1580    IF PRFLAG=1 AND PN#1 THEN PRINT FT,G,GDB
```

```
1590    DISP FT,G,GDB
1600 NEXT FT @ DISP
1610 IF PRFLAG=1 AND PN#1 THEN PRINT
1620 ON PLFLAG GOTO 650,2030
1630 !
1640 ! ***** SUB TO CALC FREQ RESPONSE OR GAIN AT SINGLE FREQ. *****
1650 X=FNXZ(r,NPTS)*COS (180*FT)
1660 M=NPTS-1 @ GOSUB 1870 @ G=TM/r @ GDB=FNVDB(G) @ RETURN
1670 !
1680 ! ********** Cheby Poly for single argument ***************
1690 DISP "Enter argument and poly order (X,M)";@ INPUT X,M
1700 IF FP (M) THEN DISP "M must be an Integer!" @ BEEP @ GOTO 1690
1710 GOSUB 1870
1720 PR$=P$&VAL$ (M)&"    T("&VAL$ (X)&")="&VAL$ (TM) @ GOSUB 2340
1730 DISP "More values";@ INPUT R$@ R$=UPC$ (R$) @ IF R$="Y" THEN 1690
1740 IF R$="N" THEN 520 ELSE DISP "Y OR N?" @ GOTO 1730
1750 !
1760 ! ************** Chebyshev Polynomial sweep **************
1770 DISP "Enter poly order, M";@ INPUT M
1780 IF FP (M) THEN DISP "M must be an Integer!" @ BEEP @ GOTO 1770
1785 IF M<1 THEN DISP "Use M>0" @ BEEP @ GOTO 1770
1790 DISP "Enter ";CHR$ (7);"x and # of points";@ INPUT DX,NPP
1795 IF DX<= 0 THEN DISP "Use ";CHR$ (7);"x>0" @ BEEP @ GOTO 1790
1797 IF NPP<1 THEN DISP "Use # of points>=1" @ BEEP @ GOTO 1790
1800 PR$="" @ GOSUB 2300 @ PR$=P$&VAL$ (M) @ GOSUB 2300
1810 PR$=Q$ @ GOSUB 2340 @ X=0
1820 FOR I=1 TO NPP @ GOSUB 1870
1830    DISP X,TM @ IF PN#1 THEN PRINT X,TM
1840    X=X+DX
1850 NEXT I @ GOTO 520
1860 !
1870  ! ************** SUB to calc  Chebyshev polynomial ********
1880 IF ABS (X)<= 1 THEN TM=COS (M*ACS (X)) ELSE TM=FNCOSH(M*FNICOSH(X))
1890 RETURN
1900 !
1910 ! ********** SUB TO CALC COSH(X) FROM KEYBOARD *****************
1920 DISP "Enter x ([EL] to Quit)";@ INPUT X$
1930 IF X$="" THEN 520 ELSE X=VAL (X$)
1935 IF ABS (X)>1148 THEN J$=")>" @ XX=1148 ELSE J$=")=" @ XX=X
1940 PR$="    COSH("&X$&J$&VAL$ (FNCOSH(XX)) @ GOSUB 2340 @ GOTO 1920
1950 !
1960 ! ********** SUB TO CALC INV COSH(X) FROM KEYBOARD ************
1970 DISP "Enter X>=1([EL] To Quit)";@ INPUT X$@ IF X$="" THEN 520
1980 X=VAL (X$) @ IF X<1 THEN DISP "Use X>=1" @ GOTO 1970
1985 IF ABS (X)>1.E249 THEN J$=")>" @ XX=1.E249 ELSE J$=")=" @ XX=X
1990 PR$="   INV COSH("&X$&J$&VAL$ (FNICOSH(XX)) @ GOSUB 2340 @ GOTO 1970
2000 !
2010 ! ************* SUB TO PLOT DFT RESPONSE **************
2030 IF PA=1 THEN LIMIT 0,125,0,75 @ X2=160 @ X3=85 @ GOTO 2050
2040 LIMIT 0,240,0,190 @ X2=120 @ X3=70
```

```
2050 GRAPH @ S$="Dolph-Chebyshev,"&VAL$ (GSL)&"dB sidelobes"
2060 FXD 2,0 @ LOCATE 15,X2,10,84 @ DEG @ GCLEAR
2070 MOVE 4,50 @ LORG 4 @ LDIR 90 @ CSIZE 5,.7,20 @ LABEL YL$
2080 N$=VAL$ (NPTS)&" Point DFT"
2090 MOVE 15,87 @ CSIZE 4 @ LDIR 0 @ LORG 3 @ LABEL N$
2100 MOVE X3,94 @ CSIZE 8 @ LORG 4 @ LABEL T$
2110 MOVE X3,89 @ CSIZE 5 @ LABEL S$
2120 MOVE X3,5 @ LORG 6 @ LABEL X$
2130 MOVE X2,5 @ CSIZE 4 @ LORG 9 @ LABEL DA$
2140 YB=-(10*INT (ABS (GSL)/10))-20
2150 SCALE 0,.5,YB,0
2160 LGRID -.01,2,0,YB,5,5
2170 AXES .01,2,.5,0,5,5,4
2180 LINE TYPE 1 @ MOVE 0,0 @ I=2
2190 FOR FT=DFT TO .5 STEP DFT
2200    DRAW FT,GGDB(I) @ I=I+1
2210 NEXT FT @ IF PA#1 THEN PEN 0 @ MOVE .5,0 @ GOTO 650
2230 ALPHA @ DISP "Plot on Plotter";@ INPUT R$
2240 R$=UPC$ (R$) @ IF R$="Y" THEN GOSUB 2620 @ GOTO 2030
2250 IF R$#"N" THEN DISP "Y OR N?" @ GOTO 2230
2255 DISP "Dump to printer";@ INPUT R$@ R$=UPC$ (R$) @ IF R$="N" THEN 650
2258 IF R$#"Y" THEN DISP "Y or N?" @ BEEP @ GOTO 2255
2259 IF PN=1 THEN GOSUB 2440 @ GOTO 2259
2260 IF PN=4 THEN DUMP GRAPHICS 0,0,0,-1 ELSE DUMP GRAPHICS
2270 GOTO 650
2280 !
2290 ! ****************  DISPLAY/PRINT& SPACE  ***************
2300 DISP PR$ @ DISP @ IF PN#1 THEN PRINT PR$ @ PRINT
2310 RETURN
2320 !
2330 ! ******************  DISPLAY/PRINT  ******************
2340 DISP PR$ @ IF PN#1 THEN PRINT PR$
2350 RETURN
2360 !
2370 DISP PR$ @ IF PRFLAG=1 THEN PRINT PR$
2380 RETURN
2390 !
2400 ! ***********  INITIALIZATION &  PRINTER SELECTION  ********
2410 PAGESIZE 24 @ CLEAR @ DEG @ PRINTER IS 1
2420 DISP TAB (10);T$ @ DISP
2430 DISP "Enter today's date (any comma-less format)";@ INPUT DA$
2440 DISP "Select Printer:" @ DISP
2450 DISP "1   CRT"
2460 DISP "2   HP-82905B--HP-IB"
2470 DISP "3   HP-82905B--Centronics"
2480 DISP "4   MX-100"
2490 DISP "Enter #";@ INPUT PN
2500 IF PN<1 OR PN>4 THEN DISP "Use # 1-4 " @ BEEP @ GOTO 2490
2505 IF PN#1 THEN DISP "Make sure Top-Of-Form is set properly, Press [END
LINE]";@ INPUT PR$
2510 ON PN GOTO 2590,2530,2540,2560
2520 !
```

```
2530 PRINTER IS 301 @ GOTO 2550
2540 PRINTER IS 701
2550 PRINT CHR$ (27)&"&11L" @ GOTO 2580 ! Skip perf.
2560 PRINTER IS 701 @ PRINT CHR$ (27);"N";CHR$ (8) ! Skip perf.
2580 PRINT TAB (15);T$ @ PRINT TAB (25);DA$
2590 DISP @ DISP "Answer questions with 'Y' or 'N', [END LINE]" @ DISP
2600 RETURN
2610 !
2620 ! ********** Sub to Select plotter, load binary GDUMP ********
2630 DISP "Enter Plotter Address (CRT=1)";@ INPUT PA
2640 IF PA<1 THEN DISP "Use # >=1" @ BEEP @ GOTO 2630
2650 ON ERROR GOTO 2660 @ GDUMP @ GOTO 2710 ! Error if GDUMP not
resident,              displays "Copyright Hewlett-Packard 203.09", if resident
2660 ON ERROR GOTO 2690 @ LOADBIN "GDUMP" @ GOTO 2650 ! several errors possible
2670 ! ERRN=67 indicates GDUMP not on disc
2680 ! ERRN=130    "       bad disc, no disc, door open, etc.
2690 IF ERRN =67 OR ERRN =130 THEN DISP "Insert disc with binary GDUMP, press
[CONT]" @ PAUSE @ GOTO 2660
2700 DISP "Unknown error,ERRN=";ERRN ;" try again" @ GOTO 2630
2710 OFF ERROR @ PLOTTER IS PA @ RETURN
```

Program PAIRECHO

```
10 ! Pgm PAIRECHO, rev 2/05/84
20 ! trans from Pgms 5-7, 5-8 of Radar Calc. by W. Skillman
30 !
40 ! Improvements vs Radar Calc.
50 !    Bandwidth ratio, weighting loss calc. on coef. entry
60 !
70 ! Important Variables:
80 !    ADB...time resp (dB)
90 !    B.....unweighted noise bandwidth
100 !   BR....bandwidth ratio
110 !   BT....normalized time
120 !   F(i)..Fourier Coef array, i=1 to N
130 !   FR....norm. Freq.
140 !   GDB...freq resp (dB)
150 !   GSL...mainlobe/sidelobe in dB, Taylor wts
160 !   LW....Wt. loss
170 !   N.....No. of Fourier Coefs.
180 !   r.....mainlobe/sidelobe voltage ratio, Taylor wts
190 !   R4....n-bar, Taylor wts
200 !   SIG...sigma, Taylor
210 !
220 DIM PR$[80],F(99),T$[26]
230 !
240 ! ***************** Defined Functions  ***************
250 DEF FNPDB(P) = 10*LGT (MAX (EPS ,P)) ! Power ratio to dB
260 DEF FNVDB(V) = FNPDB(V*V) ! Voltage ratio to dB
270 DEF FNSIG(X) = SQR (X*X/(ASQ+(X-.5)*(X-.5))) ! Sub 107, pgm 5-8
280 !
290 T$="**PAIRED ECHO ANALYSIS**"
300 GOSUB 900 ! init, printer sel.
310 GOSUB 430 ! enter Fourier coefs.
320 !
330 ! ********* Main Menu  *************
340 PR$="" @ GOSUB 560
350 DISP "Select next operation" @ DISP
360 DISP "1  New coefficients"
370 DISP "2  Time Response"
380 DISP "3  Frequency response" @ DISP "4  Disc Menu"
390 DISP @ DISP "Enter #";@ INPUT R@ IF R=4 THEN CHAIN "Autost"
400 IF R<1 OR R>4 THEN DISP "Use # 1-4" @ BEEP @ GOTO 390
410 ON R GOSUB 430,700,800 @ GOTO 340
420 !
430 ! ************ Fourier Coef Entry  ******************
440 SUMF,SUM2F=.5 @ PR$="" @ GOSUB 560
450 DISP "Select Coefs." @ DISP
460 DISP "1  User-Defined"
470 DISP "2  Taylor"
480 DISP @ DISP "Enter #";@ INPUT R
490 IF R<1 OR R>2 THEN DISP "Use 1 or 2" @ BEEP @ GOTO 480
500 ON R GOSUB 1130,1230
510 !
```

```
520 LW=2*SUM2F @ LWDB=FNPDB(LW) @ BR=LW/4/SUMF/SUMF
530 PR$="Weighting loss="&VAL$ (LWDB)&"dB, Noise bw ratio="&VAL$ (BR)
540 !
550 ! *************** display/print ****************
560 DISP PR$ @ IF PN#1 THEN PRINT PR$
570 RETURN
580 !
590 ! ********** SUB TO CALC SINC(t) **************
600 IF t=0 THEN SINC=1 @ RETURN
610 SINC=0 @ IF FP (t) THEN SINC=SIN (180*t)/PI /t
620 RETURN
630 !
640 ! ********** Sub to calc time response  at time BT *******
650 t=BT @ GOSUB 590 @ TSUM=SINC @ FOR I=1 TO N
660   t=BT+I @ GOSUB 590 @ TT=SINC @ t=BT-I @ GOSUB 590
670   TSUM=TSUM+F(I)*(TT+SINC)
680 NEXT I @ ADB=FNVDB(TSUM) @ RETURN
690 !
700 ! *************** Time Response ********************
710 DISP "Enter norm. start time, increment ";@ INPUT BT,BDT
720 IF BDT<= 0 THEN DISP "Use increment > 0" @ BEEP @ GOTO 710
730 FOR J=1 TO 10
740   GOSUB 640 ! Time Response, SINC(t)
750   PR$="Time="&VAL$ (BT)&", Response="&VAL$ (ADB) @ GOSUB 560 @ BT=BT+BDT
760 NEXT J
770 DISP "More Responses";@ INPUT R$@ R$=UPC$ (R$) @ IF R$="Y" THEN 730
780 IF R$="N" THEN RETURN ELSE DISP "Y or N?" @ BEEP @ GOTO 770
790 !
800 ! ****************** Frequency Response ***************
810 DISP "Enter norm. freq. increment";@ INPUT DF
820 IF DF<= 0 THEN DISP "Use increment > 0" @ BEEP @ GOTO 810
830 FOR FR=0 TO 180 STEP 360*DF @ FSUM=.5
840   FOR I=1 TO N
850     FSUM=FSUM+F(I)*COS (I*FR)
860   NEXT I @ GDB=FNVDB(FSUM/SUMF) @ PR$="Normalized freq.="
870   PR$=PR$&VAL$ (FR/360)&", Response="&VAL$ (GDB)&"dB" @ GOSUB 560
880 NEXT FR @ RETURN
890 !
900 ! ****** INITIALIZATION &  PRINTER SELECTION *******
910 PAGESIZE 24 @ CLEAR @ DEG @ PRINTER IS 1
920 DISP TAB (10);T$ @ DISP
930 DISP "Select Printer:" @ DISP
940 DISP "1  CRT"
950 DISP "2  HP-82905B--HP-IB"
960 DISP "3  HP-82905B--Centronics"
970 DISP "4  MX-100"
980 DISP "Enter #";@ INPUT PN
990 IF PN<1 OR PN>4 THEN DISP "Use # 1-4 " @ BEEP @ GOTO 980
995 IF PN#1 THEN DISP "Make sure Top-Of-Form is set properly, Enter
date(no',')";@ INPUT DA$
1000 ON PN GOTO 1090,1020,1030,1050
1010 !
```

```
1020 PRINTER IS 301 @ GOTO 1040
1030 PRINTER IS 701
1040 PRINT CHR$ (27)&"&11L" @ GOTO 1080
1050 PRINTER IS 701 @ PRINT CHR$ (27);"N";CHR$ (8)
1080 PRINT TAB (15);T$ @ PRINT TAB (25);DA$
1090 DISP @ DISP "Answer questions with 'Y' or 'N', [END LINE]" @ DISP
1100 RETURN
1110 !
1120 ! ********** Enter User-Defined Coefs. ***************
1130 DISP "Enter # of Fourier Coefs.";@ INPUT N
1140 IF N<1 OR N>99 THEN DISP "Use 0 < # < 100" @ BEEP @ GOTO 1130
1150 IF FP (N) THEN DISP "Use integer" @ BEEP @ GOTO 1130
1160 FOR I=1 TO N
1170   DISP "Enter F(";I;")";@ INPUT FF
1180   SUMF=SUMF+FF @ SUM2F=SUM2F+FF*FF @ F(I)=FF
1190   IF PN#1 THEN PRINT "F(";I;") = ";FF
1200 NEXT I @ RETURN
1210 !
1220 ! ************ Enter Taylor Coefficients *********
1230 GOSUB 1380 ! Enter GSL, Calc. n-bar minimum, enter n-bar
1240 SIG=FNSIG(R4) @ PR$="Sigma = "&VAL$ (SIG) @ GOSUB 560 @ R2=1
1250 FOR M=1 TO N
1260   M2=M*M @ X=.5 @ H8=M2/SIG/SIG
1270   FOR J=1 TO N
1280     X=X*(1-H8/(ASQ+(J-.5)*(J-.5)))
1290     IF J#M THEN X=X/(1-M2/J/J)
1300   NEXT J @ X=R2*X @ F(M)=X @ R2=-R2
1310   SUMF=SUMF+X @ SUM2F=SUM2F+X*X
1320   PR$="F"&VAL$ (M)&"="&VAL$ (X) @ GOSUB 560
1330 NEXT M @ X=r/SQR (2) @ GOSUB 1520
1340 X=SIG*SQR (5.0968*(ASQ-X))
1350 PR$="Widening factor= "&VAL$ (X) @ GOTO 560
1360 !
1370 ! ************ SUB B OF PGM 5-8, ENTER GSL IN DB **************
1380 DISP "Enter GSL (dB)";@ INPUT X@ GSL=ABS (X)
1390 SIG=0 @ IF PN#1 THEN PRINT "GSL =";-GSL;" dB"
1400 X,r=10^(GSL/20) @ R4=1
1410 GOSUB 1520 @ ASQ=X ! Calc A^2
1420 T=SIG @ R4=R4+1
1430 SIG=FNSIG(R4) @ IF SIG>T THEN 1420
1440 R4M=R4-1 ! Exit when sigma decreases, which defines n-bar minimum
1450 DISP "Enter n-bar>";R4M;@ INPUT R4
1460 IF R4>100 THEN DISP "Use n-bar <= 100 " @ BEEP @ GOTO 1450
1470 IF R4<R4M THEN DISP "Use n-bar > ";R4M @ BEEP @ GOTO 1450
1480 IF PN#1 THEN PRINT PR$;",  n-bar ="&VAL$ (R4)
1490 N=R4-1 @ RETURN
1500 !
1510 ! ****************** SUB B' OF PGM 5-8, CALC A^2 **********
1520 X=SQR (X*X-1)+X @ X=LOG (X)/PI @ X=X*X @ RETURN
```

Program FIRCANC

```
10 ! Pgm FIRCANC, Rev 3/01/84
20 !     Translated from Pgms 5-9 & 5-12 of Radar Calc...
30 !
40 ! Calcs resp of FIR Canceller, using stagger PRF
50 !   and also Improvement factor for same
60 !
70 ! Improvements vs. 5-9,12:
80 !   Binomial weights calc
90 !   Response Plot
100 !   Time-Varying weights each cycle or pulse
110 !   Numeric gain output
120 !
130 ! Important variables:
140 !   A = weight array A(0) to A(NPDS)
150 !   AFLAG =1, Avg over cycles, =0, single cycle
160 !   FC=FCT/Tavg
170 !   FCT=normalized freq
180 !   FLAG2: Selects ref for response: 0=Peak, 1=Noise
190 !   GGDB = Response array (dB)
200 !   IMP = Improvement factor
210 !   NPDS = no. of periods in stagger sequence
220 !   MDLS = no. of delay lines (MDLS+1 pulse canceller)
230 !   PA = Plotter address
240 !   PN = Printer address
250 !   PWFLAG = 1, Prinsen wts, =0 otherwise
260 !   REF = Ref for response
270 !   SIGT=std dev of Gaussian clutter spectrum
280 !   SUM1 = Sum of Abs values of wts
290 !   SUM2 = Sum of squares of wts
300 !   t = sum of T array, or time of pulse from ref
310 !   T = IPP array, T(1) to T(NPDS)
320 !   T360 = Tavg/360
330 !
340 DIM A(500),t(1000),T(500),PR$[60],F$[25],L$(2)[45],EN$[40],M$[25]
350 DIM X$[35],S$[40],GGDB(1024),T$[55],H$[60],NS$[30],AV$[25]
360 !
370 DEF FNPDB(X) = 10*LGT (MAX (EPS ,ABS (X)))
380 !
390 AV$="Average over Cycles"
400 SS$="Single Cycle" @ EN$="Enter normalized clutter offset and "
410 L$(1)="Peak response (square of abs sum of Ai)"
420 L$(2)="Noise (sum of squares of Ai)"
430 T$="** FIR CANCELLER RESPONSE/IMPROVEMENT FACTOR   **"
440 H$="fT                      G(VOLTAGE)          G(dB)"
450 GOSUB 2930 ! init & printer select
460 GOSUB 3150 ! Select plotter, load  GDUMP
470 T$=T$[1,25]&" **" @ D$="# of delay lines"
480 GOSUB 670 !  Select Canceller Wts
490 GOSUB 2170 ! Enter IPP sequence
500 GOSUB 2850 ! Select ref for response
510 !
```

```
520 ! ******************    MAIN MENU   ***************
530 PR$="" @ GOSUB 2760 @ DISP "MAIN MENU-- Select next operation" @ DISP
540 DISP "1  New IPPs"
550 DISP "2  New Weights"
560 DISP "3  New Reference"
570 DISP "4  Frequency Response"
580 DISP "5  Improvement Factor" @ DISP "6  Disc Menu"
590 DISP @ DISP "Enter #";@ INPUT NR@ IF NR=6 THEN CHAIN "Autost"
600 IF NR<1 OR NR>6 THEN DISP "Use # 1-6" @ BEEP @ GOTO 590
610 AFLAG=0 @ DISP @ ON NR GOTO 630,650,500,1640,1090
620 !
630 GOSUB 2170 @ GOTO 530
640 !
650 GOSUB 670 @ GOTO 530
660 !
670 ! ****** SUB TO ENTER NO. OF DELAYS AND SELECT CANC WTS  ****
680 DISP "Enter ";D$;@ INPUT MDLS
690 IF MDLS<1 OR MDLS>500 THEN DISP "Use no. 1-500" @ BEEP @ GOTO 680
700 IF PN#1 THEN PRINT D$;"=";MDLS @ PRINT
710 DISP "Select Canceller Weights" @ DISP @ PWFLAG=0
720 DISP "1 Binomial"
730 DISP "2 Prinsen Time-Varying"
740 DISP "3 User-Defined"
750 DISP @ DISP "enter #";@ INPUT R
760 IF R<1 OR R>3 THEN DISP " Use # 1-3" @ BEEP @ GOTO 750
770 A(0)=1 @ ON R GOTO 930,810,1000
780 !
790 ! ************  PRINSEN WEIGHTS  *********************
800 ! Prinsen time-varying weights are calculated on-line each pulse
810 PWFLAG=1 @ S$="Prinsen Weights" @ IF PN#1 THEN PRINT S$
820 RETURN
830 !
840 ! *************   ON-LINE PRINSEN WT CALC   *********
850 SUM1,SUM2=1
860 FOR I=1 TO MDLS @ X=1
870   FOR J=1 TO MDLS
880     IF I=J THEN 890 ELSE X=X/(1-t(I)/t(J))
890   NEXT J @ A(I)=-X @ SUM1=SUM1+ABS (X) @ SUM2=SUM2+X*X
900 NEXT I @ RETURN
910 !
920 ! ************  Calculate Binomial weights  *************
930 SUM1,SUM2,F=1
940 FOR I=1 TO MDLS
950   A(I),F=F*(1-(MDLS+1)/I) @ SUM1=SUM1+ABS (F) @ SUM2=SUM2+F*F
960 NEXT I @ S$="Binomial Weights" @ IF PN#1 THEN PRINT S$
970 RETURN
980 !
990 ! *****************   ENTER USER-DEFINED WEIGHTS   *************
1000 S$="User-defined Weights" @ IF PN#1 THEN PRINT S$
1010 IF PN#1 THEN PRINT "   Index        Weight"
1020 DISP "Enter weight for direct path = A(0)";@ INPUT A@ A(0),SUM1=A
1030 SUM2=A*A @ IF PN#1 THEN PRINT TAB (5);"0",A
1040 FOR I=1 TO MDLS
```

```
1050    DISP "Enter weight,A(";I;",";@ INPUT A@ IF PN#1 THEN PRINT TAB (5);I,A
1060    SUM1=SUM1+ABS (A) @ SUM2=SUM2+A*A @ A(I)=A
1070 NEXT I @ RETURN
1080 !
1090 ! ******************** IMP FACTOR MENU **********
1100 DISP @ DISP "Select Imp Factor Mode:" @ DISP
1110 DISP "Single Clutter Width"
1120 DISP "  1  ";SS$
1130 DISP "  2  ";AV$
1140 DISP "Clutter Width Sweep"
1150 DISP "  3  ";SS$
1160 DISP "  4  ";AV$
1170 DISP "5  Main Menu"
1180 DISP "Enter #";@ INPUT NR@ AFLAG=0 @ PR$=SS$
1190 IF NR<1 OR NR>5 THEN DISP "Use # 1-5 " @ BEEP @ GOTO 1180
1200 ON NR GOTO 1240,1230,1330,1320,530
1210 !
1220 ! ********** IMP FACT FOR SINGLE CLUTTER WIDTH **********
1230 AFLAG=1 @ PR$=AV$
1240 GOSUB 2770
1250 DISP EN$;"std dev";@ INPUT FCT,SIGT@ GOSUB 1570
1260 PR$="fcT="&VAL$ (FCT)&",  Sigma="&VAL$ (SIGT)&",  Imp fact="
1270 PR$=PR$&VAL$ (FNPDB(IMP))&" dB" @ GOSUB 2770
1280 DISP "More I's";@ INPUT R$@ R$=UPC$ (R$) @ IF R$="N" THEN 1090
1290 IF R$="Y" THEN 1250 ELSE DISP "Y or N?" @ BEEP @ GOTO 1280
1300 !
1310 ! ********** IMP FACT, CLUTTER WIDTH SWEEP *****************
1320 AFLAG=1 @ PR$=AV$
1330 GOSUB 2770 @ DISP EN$;"increment of std dev";@ INPUT FCT,DSIGT
1340 IF DSIGT=0 THEN DISP "Use increment > 0 " @ BEEP @ GOTO 1330
1350 DISP "Enter # of std devs to sweep (starts with 0)";@ INPUT NSD
1360 IF NSD<1 THEN DISP "Use # >=1" @ BEEP @ GOTO 1350
1370 PR$="Norm clutter offset="&VAL$ (FCT) @ GOSUB 2770
1380 PR$="Norm std dev       Imp Fact(dB)" @ GOSUB 2770
1390 SIGT=0 @ FOR L=1 TO NSD @ GOSUB 1570
1410    PR$=VAL$ (SIGT)&"              "&VAL$ (FNPDB(IMP)) @ GOSUB 2770
1420    SIGT=SIGT+DSIGT
1430 NEXT L @ GOTO 1090
1440 !
1450 ! ******************** SUB TO CALC IMP FACT ***************
1460 SUM4=0 @ X=PI *SIGT/TAVG @ F1=-(2*X*X) @ F2=FCT/T360
1470 IF PWFLAG THEN GOSUB 850
1480 FOR I=0 TO MDLS-1 @ SUM5=0
1490    FOR J=I+1 TO MDLS @ U=t(I)-t(J)
1500       SUM5=SUM5+A(J)*EXP (F1*U*U)*COS (F2*U)
1510    NEXT J @ SUM4=SUM4+A(I)*SUM5
1520 NEXT I @ X=1+2*SUM4/SUM2
1530 IF X THEN IMP=1/X ELSE IMP=INF /10 ! Protect against overflow, /0
1540 RETURN
1550 !
```

```
1560 ! ****************** Sub to calc Improvement Factor *********
1570 SUMG=0 @ KMAX=AFLAG*(NPDS-1)
1580 FOR K=0 TO KMAX
1590   GOSUB 1460 @ SUMG=SUMG+1/IMP @ IF AFLAG THEN GOSUB 2290
1610 NEXT K @ IMP=(KMAX+1)/SUMG @ RETURN
1620 !
1630 ! ******************** FREQ RESP MENU ****************
1640 DISP "Select Frequency Response Mode" @ DISP
1650 DISP "Single Frequency"
1660 DISP "  1  ";SS$
1670 DISP "  2  ";AV$
1680 DISP "Frequency Sweep"
1690 DISP "  3  ";SS$
1700 DISP "  4  ";AV$
1710 DISP "5 Main Menu"
1720 DISP "Enter #";@ INPUT NR@ AFLAG=0 @ PR$=SS$
1730 IF NR<1 OR NR>5 THEN DISP "Use # 1-5" @ BEEP @ GOTO 1720
1740 ON NR GOTO 1780,1770,1870,1860,530
1750 !
1760 ! ********* CALC CANC RESPONSE FOR SINGLE FREQUENCY *******
1770 AFLAG=1 @ PR$=AV$
1780 GOSUB 2770
1790 DISP "Enter normalized freq, fT";@ INPUT FT@ GOSUB 2110
1800 PR$="    fT="&VAL$ (FT)&"    G="&VAL$ (G)&"    G(dB)="&VAL$ (GDB)
1810 GOSUB 2770
1820 DISP "More freqs";@ INPUT R$@ R$=UPC$ (R$) @ IF R$="Y" THEN 1790
1830 IF R$="N" THEN 1640 ELSE DISP "Y or N?" @ GOTO 1820
1840 !
1850 ! ***************** CANCELLER RESPONSE SWEEP **************
1860 AFLAG=1
1870 DISP @ DISP "Enter freq. increment,";CHR$ (7);"fT"
1880 DISP "   (.002 min for CRT Plot, .001 for Plotter)";@ INPUT DFT
1890 PR$=T$ @ GOSUB 2770
1900 PR$=S$ @ GOSUB 2770
1910 IF AFLAG THEN PR$,M$=AV$ ELSE PR$,M$=SS$
1920 GOSUB 2770 @ PR$=H$ @ GOSUB 2770
1930 II=1 @ FOR FT=0 TO .5 STEP DFT
1940   GOSUB 2110 @ GGDB(II)=GDB @ II=II+1
1950   DISP FT,G,GDB @ IF PN#1 THEN PRINT FT,G,GDB
1960 NEXT FT
1970 DISP @ DISP "PLOT";@ INPUT R$@ R$=UPC$ (R$) @ IF R$="N" THEN 1640
1980 IF R$#"Y" THEN DISP "Y or N?" @ BEEP @ GOTO 1970
1990 GOSUB 2420
2000 GOTO 1640
2010 !
2020 ! **************** Calculate Single Cycle Gain *********
2030 SUMI,SUMQ=0 @ FF=FT/T360
2040 IF PWFLAG THEN GOSUB 850
2050 IF FLAG2 THEN REF=SUM2 ELSE REF=SUM1^2
2060 FOR I=0 TO MDLS @ PHI=FF*t(I)
2070   SUMI=SUMI+A(I)*COS (PHI) @ SUMQ=SUMQ+A(I)*SIN (PHI)
2080 NEXT I @ G=(SUMI*SUMI+SUMQ*SUMQ)/REF @ RETURN
2090 !
```

```
2100 ! *********** Calculate average gain over cycles  ********
2110 SUMG=0 @ KMAX=AFLAG*(NPDS-1)
2120 FOR K=0 TO KMAX
2130    GOSUB 2030 @ SUMG=SUMG+G @ IF AFLAG THEN GOSUB 2290
2140 NEXT K @ G=SUMG/(KMAX+1) @ GDB=FNPDB(G) @ RETURN
2150 !
2160 ! **************** ENTER IPPS  *************************
2170 DISP "Enter # of stagger IPPs";@ INPUT NPDS
2180 IF NPDS<1 THEN DISP "Use # >0" @ BEEP @ GOTO 2170
2190 t(0),SUM3=0 @ DISP "Enter IPPs in any units"
2200 FOR I=1 TO NPDS
2210    DISP "IPP(";I;")=";@ INPUT T(I)
2220    IF PN#1 THEN PRINT "IPP(";I;")=";T(I)
2230    t(I)=t(I-1)+T(I) @ SUM3=SUM3+T(I)
2240 NEXT I @ GOSUB 2350
2250 TAVG=SUM3/NPDS @ T360=TAVG/360
2260 PR$="Average IPP="&VAL$ (TAVG) @ GOTO 2770
2270 !
2280 ! ***********  SUB TO ROTATE IPPS, t's*********
2290 X=T(1) @ t(0)=0
2300 FOR I=1 TO NPDS-1
2310    T(I)=T(I+1) @ t(I)=t(I-1)+T(I)
2320 NEXT I @ T(NPDS)=X @ t(NPDS)=t(I-1)+X
2330 !
2340 ! **************  COMPLETE CALC OF t's  *********
2350 FOR I=NPDS+1 TO NPDS+MDLS-1
2360    I2=(I-NPDS) MOD NPDS @ IF I2=0 THEN I2=NPDS
2370    t(I)=t(I-1)+T(I2)
2380 NEXT I @ RETURN
2390 !
2400 ! ****************  SUB TO PLOT CANC RESPONSE **************
2420 IF PA=1 THEN LIMIT 0,125,0,75 @ X2=160 @ X3=85 @ GOTO 2440
2430 LIMIT 0,240,0,190 @ X2=120 @ X3=70
2440 YL$="Response (dB)" @ X$="Frequency*IPP=fT"
2450 NS$=VAL$ (NPDS)&" Pulse Stagger"
2460 FXD 2,0 @ LOCATE 15,X2,10,84 @ DEG @ GCLEAR
2470 MOVE 4,50 @ LORG 4 @ LDIR 90 @ CSIZE 5,.7,20 @ LABEL YL$
2480 N$=VAL$ (MDLS+1)&" Pulse Canceller"
2490 MOVE 15,87 @ CSIZE 4 @ LDIR 0 @ LORG 3 @ LABEL N$
2500 MOVE X2,87 @ LORG 9 @ LABEL NS$
2510 MOVE X3,94 @ CSIZE 8 @ LORG 4 @ LABEL T$
2520 MOVE X3,89 @ CSIZE 5 @ LABEL S$
2530 MOVE X3,5 @ LORG 6 @ LABEL X$
2540 MOVE X2,5 @ CSIZE 4 @ LORG 9 @ LABEL DA$
2550 MOVE 15,5 @ LORG 3 @ LABEL M$
2560 YB=-60 @ SCALE 0,.5,YB,0
2570 LGRID -.01,2,0,YB,5,5
2580 AXES .01,2,.5,0,5,5,4
2590 LINE TYPE 1 @ MOVE 0,GGDB(1) @ I=2
2600 FOR FT=DFT TO .5 STEP DFT
2610    DRAW FT,GGDB(I) @ I=I+1
2620 NEXT FT @ IF PA#1 THEN PEN 0 @ MOVE .5,0 @ RETURN
```

```
2630 ALPHA @ DISP "Plot on Plotter";@ INPUT R$
2640 R$=UPC$ (R$) @ IF R$="Y" THEN GOSUB 3150 @ GOTO 2400
2650 IF R$#"N" THEN DISP "Y or N?" @ BEEP @ GOTO 2630
2660 IF PN#1 THEN DISP "Dump to printer";@ INPUT R$ ELSE RETURN
2670 R$=UPC$ (R$) @ IF R$="N" THEN RETURN
2680 IF R$#"Y" THEN DISP "Y or N?" @ GOTO 2660
2685 IF PN=1 THEN GOSUB 2970 @ GOTO 2685
2690 IF PN=4 THEN DUMP GRAPHICS 0,0,0,-1 ELSE DUMP GRAPHICS
2700 RETURN
2710 !
2720 ! ***************** DISP/PRINT/SPACE ************
2730 DISP PR$ @ DISP @ IF PN#1 THEN PRINT PR$ @ PRINT
2740 RETURN
2750 !
2760 ! ***************** DISP/PRINT *************
2770 DISP PR$ @ IF PN#1 THEN PRINT PR$
2780 RETURN
2790 !
2800 ! **************** DISP /PRINT *************
2810 DISP PR$ @ IF PRFLAG=1 THEN PRINT PR$
2820 RETURN
2830 !
2840 ! ********** select ref for canc resp *********
2850 DISP "Select Reference for Canceller Response:"
2860 DISP "1   ";L$(1)
2870 DISP "2   ";L$(2)
2880 DISP "Enter #";@ INPUT R@ FLAG2=R-1
2890 IF R<1 OR R>2 THEN DISP "Use 1 or 2" @ BEEP @ GOTO 2880
2900 IF PN#1 THEN PRINT "Response Reference = ";L$(R)
2910 RETURN
2920 !
2930 ! *********** INITIALIZATION &  PRINTER SELECTION ***************
2940 PAGESIZE 24 @ CLEAR @ DEG @ PRINTER IS 1
2950 DISP TAB (10);T$ @ DISP
2960 DISP "Enter today's date (any comma-less format)";@ INPUT DA$
2970 DISP "Select Printer:" @ DISP
2980 DISP "1   CRT"
2990 DISP "2   HP-82905B--HP-IB"
3000 DISP "3   HP-82905B--Centronics"
3010 DISP "4   MX-100"
3020 DISP "Enter #";@ INPUT PN
3030 IF PN<1 OR PN>4 THEN DISP "Use # 1-4 " @ BEEP @ GOTO 3020
3035 IF PN#1 THEN DISP "Make sure Top-of-Form is set properly, Press [END
LINE]";@ INPUT PR$
3040 ON PN GOTO 3120,3060,3070,3090
3050 !
3060 PRINTER IS 301 @ GOTO 3080
3070 PRINTER IS 701
3080 PRINT CHR$ (27)&"&l1L" @ GOTO 3110
3090 PRINTER IS 701 @ PRINT CHR$ (27);"N";CHR$ (8)
3110 PRINT TAB (15);T$ @ PRINT TAB (25);DA$
3120 DISP @ DISP "Answer questions with 'Y' or 'N', [END LINE]" @ DISP
3130 RETURN
3140 !
```

```
3150 ! *********** Sub to Select plotter, load binary GDUMP **************
3160 DISP "Enter Plotter Address (CRT=1)";@ INPUT PA
3170 IF PA<1 THEN DISP "Use # >=1" @ BEEP @ GOTO 3160
3180 ON ERROR GOTO 3200 @ GDUMP @ GOTO 3250 ! Error if GDUMP not resident,
3190 !   or else displays "Copyright Hewlett-Packard 203.09", if resident
3200 ON ERROR GOTO 3230 @ LOADBIN "GDUMP" @ GOTO 3180 ! several errors possible
3210 ! ERRN=67 indicates GDUMP not on disc
3220 !    ERRN=130   "      bad disc, no disc, door open, etc.
3230 IF ERRN =67 OR ERRN =130 THEN DISP "Insert disc with binary GDUMP, press
[CONT]" @ PAUSE @ GOTO 3200
3240 DISP "Unknown error,ERRN=";ERRN ;" try again" @ GOTO 3160
3250 OFF ERROR @ PLOTTER IS PA @ RETURN
```

Program MEM

```
10 ! Pgm MEM, Rev 2/05/84
20 ! Trans from Pgm 5-10, Radar Calc.. by W. Skillman
30 !
40 ! Improvements vs Radar Calc.
50 !    All order filter coefs stored
60 !
70 ! Important Variables:
80 !    a.......filter coef. array
90 !    aa......copy of a array
100 !    as(i,j).filter coef. array, i=order, j=index
110 !    b1,b2...intermediate arrays
120 !    M.......Filter order
130 !    MF......Order of filter for spectrum calc.
140 !    MM......Max order of filter
150 !    N.......no. of input data points
160 !    P.......Filter output power
170 !    X.......Input data (time series)
180 !
190 DIM PR$[80],T$[28],a(512),aa(512),b1(512),b2(512),as(50,50)
200 !
210 T$="**Maximum Entropy Method**"
220 !
230 ! **************** Defined Functions *******************
240 DEF FNPDB(P) = 10*LGT (MAX (EPS ,P)) ! Power ratio to dB
250 !
260 GOSUB 1000 ! init, printer sel
270 GOSUB 500 ! Enter data
280 GOSUB 810 ! Select filter order, calc. spectrum
290 !
300 ! ****************** Main Menu ****************************
310 PR$="" @ GOSUB 420 @ DISP "Select next operation"
320 DISP @ DISP "1  New input data (time series)"
330 DISP "2  MEM Spectrum of New Order Filter" @ DISP "3  Disc Menu"
340 DISP @ DISP "Enter #";@ INPUT R@ IF R=3 THEN CHAIN "Autost"
350 IF R<1 OR R>3 THEN DISP "Use #  1-3 " @ BEEP @ GOTO 340
360 ON R GOTO 270,280
370 !
380 ! ******** sub to print filter coefs, store in as array ******
390 PR$="    a("&VAL$ (t)&")="&VAL$ (a(t)) @ IF M<51 THEN as(M,t)=a(t)
400 !
410 ! ************* disp/print *****************************
420 DISP PR$ @ IF PN#1 THEN PRINT PR$
430 RETURN
440 !
450 ! *********** Sub to enter data point ***************
460 DISP "Enter X(";t;")";@ INPUT X
470 IF PN#1 THEN PRINT "X(";t;")=";X
480 SUM2=SUM2+X*X @ b1(t),b2(t-1)=X @ RETURN
490 !
```

```
500 ! ***********  Enter Input data   ************************
510 DISP "Enter # of data points";@ INPUT N
530 IF N>512 OR N<= 0 THEN DISP "Use 0 < # < 513" @ BEEP @ GOTO 510
540 SUM2=0 @ FOR t=1 TO N @ GOSUB 450 @ NEXT t
550 DISP "Any corrections";@ INPUT R$@ R$=UPC$ (R$)
560 IF R$="N" THEN 630
570 IF R$#"Y" THEN DISP "Y or N?" @ BEEP @ GOTO 550
580 DISP "Enter index of data (0 to quit)";@ INPUT t
590 IF t>N OR t<0 THEN DISP "Use 0<index<=";N @ BEEP @ GOTO 580
600 IF t#0 THEN SUM2=SUM2-b1(t)*b1(t) @ GOSUB 450 @ GOTO 580
610 !
620 ! ****************  Calc Coefs    *************************
630 P=SUM2/N @ M=1 @ PR$="P(0)="&VAL$ (P) @ GOSUB 420
640 DISP "Enter max order filter (default=";N-1;")";@ INPUT MM$
650 IF MM$="" THEN MM=N-1 ELSE MM=VAL (MM$)
660 IF MM<1 THEN DISP "Use order>=1" @ BEEP @ GOTO 640
670 NOM,DEN=0 @ FOR t=1 TO N-M
680    NOM=NOM+b1(t)*b2(t) @ DEN=DEN+b1(t)*b1(t)+b2(t)*b2(t)
690 NEXT t @ am,a(M)=2*NOM/DEN @ M$=VAL$ (M) @ P=P*MAX (0,1-am*am)
700 PR$="Filter order="&M$ @ GOSUB 420
710 PR$="   P("&M$&")="&VAL$ (P) @ GOSUB 420
720 IF M=1 THEN t=1 @ GOTO 750
730 FOR t=1 TO M-1 @ a(t)=aa(t)-am*aa(M-t) @ GOSUB 390 @ NEXT t
740 !
750 GOSUB 390 @ IF M=MM OR P<= 0 THEN MM=M @ RETURN
760 M=M+1 @ FOR t=1 TO M-1 @ aa(t)=a(t) @ NEXT t
770 FOR t=1 TO N-M
780    b1(t)=b1(t)-am*b2(t) @ b2(t)=b2(t+1)-am*b1(t+1)
790 NEXT t @ GOTO 670
800 !
810 ! ***********  DFT of  Filter Coefficients  *****************
820 DISP "Select order of filter (<=";MM;")";@ INPUT MF
830 IF MF<1 OR MF>MM THEN DISP "Use 0<order<";MM @ BEEP @ GOTO 820
840 PR$="Filter order="&VAL$ (MF) @ GOSUB 420
850 DISP "Enter min, max, and increment of norm freq (0-.5)";
860 INPUT F1,F2,DF@ IF F2<F1 THEN DISP "Use max>min" @ BEEP @ GOTO 850
870 IF DF<= 0 THEN DISP "Use increment >0" @ BEEP @ GOTO 850
880 FOR FT=F1 TO F2 STEP DF @ GOSUB 930
890    PR$="Freq="&VAL$ (FT)&",  Power ="&VAL$ (GDB) @ GOSUB 420
900 NEXT FT @ DISP "More freqs";@ INPUT R$@ R$=UPC$ (R$)
910 IF R$="Y" THEN 850 ELSE RETURN
920 !
930 ! ************  Sub to calc single freq response  ************
940 FF=360*FT @ CS=2*COS (FF) @ D1,D2=0
950 FOR I=1 TO MF
960    T=D1 @ D1=as(MF,I)-D2+CS*D1 @ D2=T
970 NEXT I @ RE=1-D1*CS/2+D2 @ IM=D1*SIN (FF) @ P=RE*RE+IM*IM
980 P=MAX (EPS ,P) @ GDB=FNPDB(1/P) @ RETURN
990 !
```

```
1000 ! ***********  INITIALIZATION &  PRINTER SELECTION  ***********
1010 PAGESIZE 24 @ CLEAR @ DEG @ PRINTER IS 1
1020 DISP TAB (10);T$ @ DISP
1030 DISP "Select Printer:" @ DISP
1040 DISP "1   CRT"
1050 DISP "2   HP-82905B--HP-IB"
1060 DISP "3   HP-82905B--Centronics"
1070 DISP "4   MX-100"
1080 DISP "Enter #";@ INPUT PN
1090 IF PN<1 OR PN>4 THEN DISP "Use # 1-4 " @ BEEP @ GOTO 1080
1095 IF PN#1 THEN DISP "Make sure Top-Of-Form is set properly, Enter
date(no',')";@ INPUT DA$
1100 ON PN GOTO 1190,1120,1130,1150
1110 !
1120 PRINTER IS 301 @ GOTO 1140
1130 PRINTER IS 701
1140 PRINT CHR$ (27)&"&l1L" @ GOTO 1180
1150 PRINTER IS 701 @ PRINT CHR$ (27);"N";CHR$ (8)
1180 PRINT TAB (15);T$ @ PRINT TAB (25);DA$
1190 DISP @ DISP "Answer questions with 'Y' or 'N', [END LINE]" @ DISP
1200 RETURN
```

Program IMPFACT

```
10 ! Pgm IMPFACT, Rev 2/29/84 ,written by W. Skillman
20 ! Trans from Pgm  5-11 of "Radar Calc...
30 ! Calculates Improvement Factor for DFT "Filter Bank"
40 !    Real Weights are assumed positive,  with even symmetry
50 !    Complex wts are in I,Q format for specific filters
60 !
70 ! Improvements vs. Radar Calc...
80 !   Non-coherent detector added
90 !   Uniform weighting or Dolph-Chebyshev loaded automatically
100 !   CRT or plotter plot, dump to printer from CRT.
110 !   Complex wts permit arbitrary filter positioning
120 !
130 ! PROGRAM FEATURES:
140 !    Coherent or non-coherent detector
150 !    Weighting options:
160 !       User-defined, Uniform, Dolph-Chebyshev, or Hamming
170 !    Gaussian clutter spectrum
180 !
190 ! Important Variables:
200 !     AI = Inphase component of complex weight
210 !     AM = weight amplitude array
220 !     AQ = Quadrature comp. of cplx wt
230 !     CFLAG = 1, Complex wts, =0, Real wts
240 !     DFREQT= frequency step for sweep
250 !     FF = output I array
260 !     FREQT = frequency/PRF
270 !     GAMN = Clutter spectrum amplitude factor
280 !     NCO = 1, coherent det, =2 non-coh det
290 !     NWT   = # of points in DFT
300 !     ODDFLAG = 0, NWT even, =1, NWT odd
310 !     PA,PN = Plotter and Printer addresses
320 !     PFLAG = print flag
330 !     PH = weight phase array
340 !     SIGT  = standard deviation of Gaussian clutter spectrum/PRF
350 !     SUMA = sq. root of num. of I, or sum AM
360 !     SUMB = denominator of I
370 !     WT(I) =  Real DFT Weights , 1<=I<=NWT
380 !
390 DIM WT(1024),FF(301),S$[40],DC$[25],DMN(1023),CC(1023),PR$[70]
400 DIM DD$[30],T$[35],HDR$[65],X$[30],YL$[25],YR$[25],DETT$[25]
410 DIM AI(200),AQ(200),AM(200),PH(200),B(200)
420 !
430 INTEGER I,IMAX,J,K,L,M,N,NCO,NMAX,NWT,ODDFLAG,PFLAG,PA,PN,R,X2,X3
440 !
450 NP$=" point DFT"
460 DC$="Sidelobe level in dB" @ DA$=""
470 T$="**DFT IMPROVEMENT FACTOR**" @ B$="Sigma/PRF" @ WT$="Weights"
480 HDR$="Index            Freq/PRF               I(dB)"
490 CC$="Coherent"
```

```
500 X$="Frequency/PRF"
510 YL$="Improvement Factor (dB)"
520 YR$="Clutter (dB)" @ DT$=" Det."
530 GOSUB 3120 @ GOSUB 3340 @ GOTO 2620 ! init + select real vs complex wts
540 !
550 DISP "DFT ";WT$;" are assumed symmetrical" @ DISP
560 GOSUB 830 @ GOSUB 1520 ! Select Detector and Sigma of clutter
570 GOSUB 770 ! # of points in DFT
580 GOSUB 1560 ! Select weights
590 !
600 ! ******************** MAIN MENU ***********************
610 PR$="" @ GOSUB 2590 @ DISP "Main menu: select next operation" @ DISP
620 DISP "1  New # of points in DFT"
630 DISP "2  New weights"
640 DISP "3  New detector"
650 DISP "4  New sigma/PRF"
660 DISP "Improvement Factor Calc."
670 DISP "  5  Single freq"
680 DISP "  Sweep:"
690 DISP "  6  Print"
700 DISP "  7  Plot"
710 DISP "  8  Plot&Print" @ DISP "9  Disc Menu"
720 DISP @ DISP "Enter #";@ INPUT R@ IF R=9 THEN CHAIN "Autost"
730 IF R<1 OR R>9 THEN DISP "Use 1-9" @ BEEP @ GOTO 720
740 ON R GOTO 570,580,1130,1490,1160,1210,1320,1315
750 !
760 ! *********** ENTER NO. OF POINTS IN DFT ******************
770 DISP "Enter # of points in DFT ";@ INPUT NWT
780 IF NWT<2 OR NWT>1024 THEN DISP "Use # 2-1024" @ BEEP @ GOTO 770
790 IF PN#1 THEN PRINT NWT;NP$
800 NMAX=INT (NWT/2) @ ODDFLAG=2*FP (NWT/2) @ RETURN
810 !
820 ! *************** SELECT TYPE OF DETECTOR ****************
830 DISP @ DISP "Select type of detector" @ DISP
840 DISP "1  Coherent detector"
850 DISP "2  Non-coherent detector"
860 DISP @ DISP "Enter #";@ INPUT R
870 IF R<1 OR R>2 THEN DISP "Use  1 or 2" @ BEEP @ GOTO 860
880 IF R=1 THEN GAM1=1 @ NCO=1 @ PR$="" ELSE GAM1=.915 @ NCO=2 @ PR$="Non-"
890 PR$,DETT$=PR$&CC$&DT$ @ GOTO 2600
900 !
910 ! ********************** IMPROVEMENT FACTOR **************
920 ! Improvement factor for single freq., input= FREQT, output=IMP
930 SUMB=0 @ GAMN=GAM1 @ L=1
940 FOR J=1 TO NWT-1 @ CC(J)=COS (360*FREQT*J) @ NEXT J
950 LL=-(2*L*NCO*PI *PI *SIGT*SIGT)
960   FOR J=1 TO NWT-1 @ DMN(J)=EXP (LL*J*J)*CC(J) @ NEXT J
970   IF ODDFLAG THEN SUMD=.5*WT(NMAX+1)^2 ELSE SUMD=0
980   FOR N=1 TO NMAX @ SUMC=0
990     FOR M=N+1 TO NMAX
1000       SUMC=SUMC+WT(M)*(DMN(M-N)+DMN(NWT-M+1-N))
1010     NEXT M @ IF ODDFLAG THEN SUMC=SUMC+WT(M)*DMN(M-N)
1020     SUMD=(2*SUMC+WT(N)*(1+DMN(NWT-2*N+1)))*WT(N)+SUMD
```

```
1030    NEXT N @ SUMD=ABS (2*SUMD*GAMN) @ SUMB=SUMB+SUMD
1040    IF SUMB THEN R=SUMD/SUMB ELSE R=1
1050    IF R<= .0001 OR NCO=1 THEN 1090 ! Exit if converged or coh det
1060    L=L+1 @ X=1-1.5/L @ GAMN=GAMN*X*X ! New spectrum for non-coh det
1070 GOTO 950
1080 !
1090 IF SUMB THEN IMP=10*LGT (SUMA*SUMA/SUMB) ELSE IMP=500
1100 RETURN
1110 !
1120 ! *****************   SELECT DETECTOR   **************
1130 GOSUB 830 @ GOTO 610
1140 !
1150 ! ******************  SINGLE FREQ  IMP FACT  *********
1160 DISP "Enter freq/PRF";@ INPUT FREQT@ GOSUB 920
1170 PR$="freq/PRF="&VAL$ (FREQT)&"   IMP FACT (dB)="&VAL$ (IMP)
1180 GOSUB 2560 @ GOTO 610
1190 !
1200 ! ****************   IMP FACT SWEEP   *******************
1210 PFLAG=1
1220 DISP "Enter ";CHR$ (7);"freq/PRF";@ INPUT DFREQT
1230 DISP "Enter no. of freqs, (sweep starts at 0)";
1240 INPUT NF@ PR$=HDR$ @ GOSUB 2600 @ I,FREQT=0
1250 GOSUB 1430 @ FREQT=FREQT+DFREQT @ IF I<NF THEN 1250 ELSE 610
1300 !
1310 ! *************   PRINT/PLOT IMP FACT   ********************
1315 PFLAG=1 @ GOTO 1330
1320 PFLAG=0
1330 DISP HDR$ @ IF PFLAG=1 AND PN#1 THEN PRINT HDR$
1340 GOSUB 1410 @ X$="Frequency/PRF" @ CFLAG=0 @ XR=.5 @ DELX=.05
1350 NX=5 @ XP=2 @ GOSUB 2180 @ GOTO 610
1390 !
1400 ! ****************   SWEEP FOR PLOT   ******************
1410 DFREQT=.01 @ I=0
1420 FOR FREQT=0 TO .49 STEP DFREQT @ GOSUB 1430 @ NEXT FREQT
1430 I=I+1 @ GOSUB 920 @ DISP I,FREQT,IMP
1450 IF PFLAG=1 AND PN#1 THEN PRINT I,FREQT,IMP
1460 FF(I)=IMP @ RETURN
1470 !
1480 ! ***************   NEW SIGMA/PRF   ********************
1490 GOSUB 1520 @ GOTO 610
1500 !
1510 ! ***************   NEW SIGT   ********************
1520 DISP "Enter ";B$;@ INPUT SIGT@ IF PN#1 THEN PRINT B$;"=";SIGT
1530 RETURN
1540 !
```

```
1550 ! ********************  WEIGHTING  ********************
1560 DISP @ DISP "SELECT TYPE OF WEIGHTING" @ DISP
1570 DISP "1   User-defined"
1580 DISP "2   Hamming"
1590 DISP "3   Dolph-Chebyshev"
1600 DISP "4   Uniform"
1610 SUMA=0 @ IMAX=INT (NWT/2+.5) @ YT=60
1620 DISP @ DISP "Enter #";@ INPUT R
1630 IF R<1 OR R>4 THEN DISP "Use # 1-4" @ BEEP @ GOTO 1620
1640 ON R GOTO 1670,1780,1900,1860
1650 !
1660 ! **********  USER-DEFINED WEIGHTS  *******************
1670 IF PN#1 THEN PRINT @ PRINT "User-defined weights" @ PRINT
1680 DISP "Enter name for weights (<40 char, no commas)";@ INPUT S$
1690 DISP "Enter max I (dB) (multiple of 10) for plot";@ INPUT R
1700 YT=ABS (R) @ IF PN#1 THEN PRINT "   I             WT(I)"
1710 FOR I=1 TO IMAX
1720   DISP "Enter WT(";I;")";@ INPUT WT(I)@ IF PN#1 THEN PRINT I,WT(I)
1730   SUMA=SUMA+WT(I)
1740 NEXT I
1750 SUMA=SUMA*2-ODDFLAG*WT(IMAX) @ RETURN
1760 !
1770 ! *******************  HAMMING WEIGHTS  *****************
1780 ANG2=360/NWT @ ANG=ANG2/2
1790 PR$,S$="Hamming "&WT$ @ GOSUB 2560
1800 FOR I=1 TO IMAX
1810   WT(I)=1-.8575*COS (ANG) ! Note: more accurate than Pgm 5-11
1820   ANG=ANG+ANG2 @ SUMA=SUMA+WT(I)
1830 NEXT I @ GOTO 1750
1840 !
1850 ! *****************  UNIFORM WEIGHTS  *******************
1860 PR$,S$="Uniform "&WT$ @ GOSUB 2560
1870 FOR I=1 TO IMAX @ WT(I)=1 @ NEXT I @ SUMA=NWT @ RETURN
1880 !
1890 ! ***************  DOLPH-CHEBYSHEV WEIGHTS  *************
1900 PR$,S$="Dolph-Chebyshev "&WT$ @ GOSUB 2560 @ DISP "Enter ";DC$;
1910 INPUT GSL@ GSL=ABS (GSL) @ S$=S$&": -"&VAL$ (GSL)&"dB"
1920 YT=20+10*INT (GSL/10)
1930 IF PN#1 THEN PRINT TAB (5);DC$;"= -";GSL
1940 R=10^(GSL/20) @ N1=NWT-1
1950 X=(R+SQR (R*R-1))^(2/N1)
1960 R4=(X-1)/(X+1) @ R4=R4*R4 @ XN=R4*N1
1970 KK,K=INT (NWT/2+.5) @ J=1
1980 GOSUB 2100 @ XN=XN/X
1990 X=1/X @ GOSUB 2060
2000 IF K=2 THEN 2040
2010 X=XN @ GOSUB 2060
2020 IF K=3 THEN 2040
2030 FOR K=3 TO KK-1 @ GOSUB 2100 @ GOSUB 2060 @ NEXT K
2040 X=1 @ GOSUB 2060 @ GOTO 1750
2050 !
```

```
2060 ! ******************** SUB TO STORE WTS   ***************
2070 WT(J)=X @ SUMA=SUMA+X @ J=J+1 @ RETURN
2080 !
2090 ! *************** SUB TO CALC ONE COEF.   **************
2100 R10,X=1 @ FOR I=1 TO K-2
2120    X=X*R4*(K-1-I)*(NWT-K-I)/(I+I*I) @ R10=R10+X
2130    IF ABS (X)/R10<= 1.E-13 THEN 2150
2140 NEXT I
2150 X=R10*XN @ RETURN
2160 !
2170 ! *************** SUB TO PLOT DATA   ***********************
2180 DEG @ IF PA#1 THEN 2200
2190 LIMIT 0,125,0,75 @ X2=150 @ X3=85 @ GRAPH @ GCLEAR @ GOTO 2210
2200 LIMIT 0,240,0,190 @ X2=110 @ X3=70
2210 LOCATE 13,X2,10,84 @ GOSUB 2420 @ SETGU
2220 MOVE X3,89 @ CSIZE 5 @ LDIR 0 @ LORG 4 @ LABEL S$
2230 MOVE 10,87 @ CSIZE 4 @ LORG 3 @ LABEL NWT;NP$
2240 MOVE 13,5 @ LABEL DETT$
2250 MOVE X2,87 @ LORG 9 @ IF CFLAG=0 THEN LABEL B$,"=",SIGT
2260 MOVE X2,5 @ LABEL DA$ @ SETUU
2270 MOVE 0,FF(1) @ I=1
2280 FOR FREQT=DFREQT TO XR STEP DFREQT
2290    I=I+1 @ DRAW FREQT,FF(I)
2300 NEXT FREQT @ IF SIGT=0 OR CFLAG THEN 2350
2310 MOVE 0,YT @ FREQT=0 @ LINE TYPE 7
2320 FREQT=FREQT+DFREQT @ Y=FREQT/SIGT
2330    Y=YT-Y*Y*2.1715 @ DRAW FREQT,Y
2340 IF Y>0 AND FREQT<.5 THEN 2320
2350 ALPHA @ IF PA#1 THEN PEN 0 @ MOVE 1,0 @ RETURN
2360 DISP "Dump to Printer";@ INPUT R$@ R$=UPC$ (R$) @ IF R$="N" THEN RETURN
2380 IF R$#"Y" THEN DISP "Y or N?" @ BEEP @ GOTO 2360
2390 IF PN=1 THEN GOSUB 3160 @ GOTO 2390
2400 IF PN=4 THEN DUMP GRAPHICS 0,0,0,-1 @ RETURN ELSE DUMP GRAPHICS @ RETURN
2410 !
2420 FXD XP,0 @ MOVE 4,50 @ LORG 4 @ LDIR 90 @ CSIZE 5,.7,30 @ LABEL YL$
2430 MOVE X3,94 @ CSIZE 8 @ LDIR 0 @ LABEL T$
2440 MOVE X3,5 @ LORG 6 @ CSIZE 5 @ LABEL X$
2450 DELY=1 @ IF YT>50 THEN DELY=2
2460 IF YT>100 THEN DELY=4
2470 SCALE 0,XR,0,YT @ CSIZE 4
2480 FRAME @ LGRID -(DELX/5),DELY,0,0,5,5
2490 XAXIS YT,-(DELX/5) @ YAXIS XR,-DELY,0,160 @ LORG 2 @ II=0
2500 IF CFLAG THEN RETURN
2510 YY=YT+II @ MOVE .505,YY @ LABEL II @ II=II-5*DELY
2520 IF ABS (II)<= YT THEN 2510
2530 MOVE .54,YT/2 @ LORG 5 @ CSIZE 5,.7,30 @ LDIR 90 @ LABEL YR$ @ RETURN
2540 !
2550 ! ******************** DISPLAY, PRINT, SPACE ***************
2560 DISP PR$ @ DISP @ IF PN#1 THEN PRINT PR$ @ PRINT
2570 RETURN
2580 !
```

```
2590 ! ********************* DISPLAY AND PRINT *******************
2600 DISP PR$ @ IF PN#1 THEN PRINT PR$
2610 RETURN
2620 !
2630 ! **************** Select Real or Complex Weights *****************
2640 DISP "Select type of DFT weights" @ DISP
2650 DISP "1  Real, symmetrical, positive wts, I vs Doppler, clut std dev."
2660 DISP "2  Complex weights, I vs clut Std Dev only"
2670 DISP @ DISP "Enter #";@ INPUT R@ IF R=1 THEN 550
2680 IF R<1 OR R>2 THEN DISP "Use 1 or 2" @ BEEP @ GOTO 2670
2690 !
2700 ! ************* Complex Weights ****************************
2710 ! Weights for Example Radar, 1st filter
2720 ! Other weights can be put into program at any unused location,
2730 !    change RESTORE address to correspond to start of DATA.
2740 !    Format of data: NWT, I(1), Q(1)...I(NWT),Q(NWT).
2750 !    COMPLEX COEFFICIENTS.  NO DOPPLER SHIFT, WEIGHTS NOT SYM.
2760 DATA 36,-1,-10,-2,-3,-3,4,-2,6,0,8,3,7,4,6,5,2,4,0,3,-4,3,-5,2,-7
2770 DATA -1,-8,-3,-8,-5,-5,-6,-5,-5,-2,-5,0,-5,4,-4,6,0,7,2,7,3,8
2780 DATA 4,6,7,4,6,2,5,0,3,-4,2,-5,-2,-6,-3,-6,-6,-6,-6,-4,-4,-1,1,4,6,8
2790 T$="**COMPLEX IMPROVEMENT FACTOR**"
2800 SUMA=0 @ K=1 @ RESTORE 2760 @ READ NWT
2810 FOR I=1 TO NWT ! Read data and calc. numerator of I
2820    READ AI(I),AQ(I)
2830    AM(I)=SQR (AI(I)*AI(I)+AQ(I)*AQ(I))
2840    PH(I)=ATN2 (AQ(I),AI(I)) @ SUMA=SUMA+AM(I)
2850 NEXT I
2860 DISP "Enter Increment of norm. std. dev. of clutter";@ INPUT DSIGT
2870 DISP "Enter # of values of I(<302), starting at std. dev. =0";@ INPUT NI
2880 FOR II=1 TO NI @ SIGT=(II-1)*DSIGT
2890    CC=EXP (-(2*(PI *SIGT)^2))
2900    FOR I=1 TO NWT-1 @ B(I)=CC^(I*I) @ NEXT I @  SUMII=0
2920    FOR n=1 TO NWT @ SUMI=AI(n)/2 @ SUMQ=AQ(n)/2
2930      FOR m=n+1 TO NWT ! Use symmetry to do 1/2 of matrix
2940        BMN=B(ABS (n-m))
2950        SUMI=SUMI+BMN*AI(m)
2960        SUMQ=SUMQ+BMN*AQ(m)
2970      NEXT m
2980      AM=AM(n)*SQR (SUMI^2+SUMQ^2)
2990      PH=PH(n)-ATN2 (SUMQ,SUMI)
3000      SUMII=SUMII+AM*COS (PH)
3010    NEXT n @ SUMB=2*SUMII @ GOSUB 1090
3020    PR$="Std. dev="&VAL$ (SIGT)&"      Imp Fact (dB)="&VAL$ (IMP)
3030    GOSUB 2600 @ FF(K)=IMP @ K=K+1
3040 NEXT II
3050 DISP "PLOT";@ INPUT R$@ R$=UPC$ (R$) @ IF R$="N" THEN 2640
3060 IF R$#"Y" THEN DISP "Y or N?" @ BEEP @ GOTO 3050
3070 X$="Clut. Std. Dev./PRF" @ CFLAG=1 @ XR=DSIGT*(NI-1)
3080 YT=10*(INT (FF(1)/10)+1) @ YB=10*INT (FF(K-1)) @ S$="" @ DETT$=CC$&DT$
3090 XP=-INT (LGT (DSIGT))
3100 DFREQT=DSIGT @ DELX=DSIGT @ NX=1 @ GOSUB 2180 @ GOTO 2640
3110 !
```

```
3120 ! ***********  INITIALIZATION &  PRINTER SELECTION  ***************
3130 PAGESIZE 24 @ CLEAR @ DEG @ PRINTER IS 1
3140 DISP TAB (10);T$ @ DISP
3150 DISP "Enter today's date (any comma-less format)";@ INPUT DA$
3160 DISP "Select Printer:" @ DISP
3170 DISP "1  CRT"
3180 DISP "2  HP-82905B--HP-IB"
3190 DISP "3  HP-82905B--Centronics"
3200 DISP "4  MX-100"
3210 DISP "Enter #";@ INPUT PN
3220 IF PN<1 OR PN>4 THEN DISP "Use # 1-4 " @ BEEP @ GOTO 3210
3225 IF PN#1 THEN DISP "Make sure Top-Of-Form is set properly, Press [END
LINE]";@ INPUT PR$
3230 ON PN GOTO 3310,3250,3260,3280
3240 !
3250 PRINTER IS 301 @ GOTO 3270
3260 PRINTER IS 701
3270 PRINT CHR$ (27)&"&l1L" @ GOTO 3300
3280 PRINTER IS 701 @ PRINT CHR$ (27);"N";CHR$ (8)
3300 PRINT TAB (15);T$ @ PRINT TAB (25);DA$
3310 DISP @ DISP "Answer questions with 'Y' or 'N', [END LINE]" @ DISP
3320 RETURN
3330 !
3340 ! ********** Sub to Select plotter, load binary GDUMP **************
3350 DISP "Enter Plotter Address (CRT=1)";@ INPUT PA
3360 IF PA<1 THEN DISP "Use # >=1" @ BEEP @ GOTO 3350
3370 ON ERROR GOTO 3380 @ GDUMP @ GOTO 3430 ! Error if GDUMP not
resident,          displays "Copyright Hewlett-Packard 203.09", if resident
3380 ON ERROR GOTO 3410 @ LOADBIN "GDUMP" @ GOTO 3370 ! several errors possible
3390 ! ERRN=67 indicates GDUMP not on disc
3400 ! ERRN=130    "        bad disc, no disc, door open, etc.
3410 IF ERRN =67 OR ERRN =130 THEN DISP "Insert disc with binary GDUMP, press
[CONT]" @ PAUSE @ GOTO 3380
3420 DISP "Unknown error,ERRN=";ERRN ;" try again" @ GOTO 3350
3430 OFF ERROR @ PLOTTER IS PA @ RETURN
```

Chapter 6 Receivers and Processing

Program NOISEFIG

```
10 ! Pgm NOISEFIG, Rev 2/11/84
20 ! Written by W. SKILLMAN, derived from PGM 6-1 OF "RADAR CALC...
30 !
40 ! Important Variables (temperatures in degrees Kelvin)
50 !    ALPH...attenuator setting, 3 dB method
60 !    ENR....Excess Noise Ratio of hot source
70 !    LR.....transmission line loss (num), LRDB in dB
80 !    NF....noise figure
90 !    PN....printer code
100 !    TA....antenna temp.
110 !    TC....temp of cold source
120 !    TE....receiver noise temp
130 !    TNS...temp of hot source
140 !    TRF...transmission line temp
150 !    Y.....Y-factor
160 !
170 DIM T$[40],PR$[90],CS$[40]
180 !
190 ! **************** Defined Functions ****************
200 DEF FNPDB(X) = 10*LGT (X) !  Power ratio to dB
210 DEF FNIPDB(X) = 10^(X/10) ! dB to power ratio
220 !
230 T$="NOISE FIGURE CALCULATIONS" @ Q$="(0 to quit)"
240 GOSUB 1140 ! Init, printer selection
250 !
260 PR$="" @ GOSUB 1110 @ DISP "Select operation:" @ DISP
270 DISP "1  Noise figure to Te"
280 DISP "2  Te to Noise figure"
290 DISP "3  System Te and noise figure"
300 DISP "4  Noise figure-cascaded stages"
310 DISP "5  Noise figure meas." @ DISP "6  Disc Menu"
320 DISP @ DISP "Enter #";@ INPUT R@ IF R=6 THEN CHAIN "Autost"
330 IF R<1 OR R>6 THEN DISP "Use # 1-6" @ BEEP @ GOTO 320
340 ON R GOTO 360,430,500,620,740
350 !
360 ! ******************** NOISE FIG TO Te  ****************
370 DISP "Enter Noise figure (dB)";Q$;
380 INPUT NF@ IF NF=0 THEN 260
390 TE=290*(FNIPDB(NF)-1)
400 PR$="Noise fig.="&VAL$ (NF)&" dB,   Te ="&VAL$ (TE)&" deg K"
410 GOSUB 1100 @ GOTO 370
420 !
430 ! ****************** Te TO NOISE FIG  ****************
440 DISP "Enter Te(deg K)";Q$;@ INPUT TE
450 IF TE<= 0 THEN 260
460 NF=1+TE/290 @ NF=FNPDB(NF)
470 PR$="Te ="&VAL$ (TE)&"deg K,   Noise fig ="&VAL$ (NF)&"dB"
480 GOSUB 1100 @ GOTO 440
490 !
```

```
500 ! *********************** SYSTEM NOISE TEMP *****************
510 DISP "Enter Antenna noise temp, Ta (deg K)";Q$;@ INPUT TA
520 IF TA=0 THEN 260
530 DISP "Enter trans. line temp(deg K) and loss(dB)";@ INPUT TRF,LRDB
540 DISP "Enter Te of rec.";@ INPUT TE
550 LRDB=ABS (LRDB) @ LR=FNIPDB(LRDB)
560 TS=TA+LR*TE+TRF*(LR-1) @ FS=FNPDB(TS/290)
570 PR$="Ta="&VAL$ (TA)&"deg K ,    TRF="&VAL$ (TRF)&",    LR="&VAL$ (LRDB)
580 PR$=PR$&"dB,  Te="&VAL$ (TE)&"deg K" @ GOSUB 1100
590 PR$="Ts="&VAL$ (TS)&" deg K,    System noise factor ="&VAL$ (FS)&" dB"
600 GOSUB 1100 @ GOTO 510
610 !
620 ! ******************** CASCADED STAGE NF ***********************
630 GIOLD,GT,I=1 @ NFOLD=0 @ PR$=" stage              Noise fig              "
640 PR$=PR$&"Gain(dB)        total Noise fig(dB)" @ GOSUB 1110
650 DISP "Enter Noise figure (dB)";Q$;", and gain(dB) of stage ";I;
660 INPUT NFIDB,GIDB@ IF NFIDB=0 THEN 260
670 GI=FNIPDB(GIDB) @ GT=GT*GIOLD @ NFI=FNIPDB(NFIDB)
680 IF I=1 THEN NFT=NFI ELSE NFT=NFT+(NFI-1)/GT
690 NFTDB=FNPDB(NFT)
700 DISP I,NFIDB,GIDB,NFTDB @ IF PN#1 THEN PRINT I,NFIDB,GIDB,NFTDB
710 I=I+1 @ GIOLD=GI @ GOTO 650
720 !
730 ! ********************** Noise Figure Measurement *************
740 DISP "Enter temp. of cold source (deg K )(290 DEFAULT)";@ INPUT TC$
750 IF TC$="" THEN TC=290 ELSE TC=VAL (TC$)
760 CS$=",   Cold Source temp ="&VAL$(TC)&" deg K"
770 DISP "Select Hot Noise Source Input:"
780 DISP @ DISP "1  Noise temperature"
790 DISP "2  Excess Noise Ratio"
800 DISP "Enter #";@ INPUT R@ IF R=2 THEN 880
810 IF R#1 THEN DISP "Use 1 or 2" @ BEEP @ GOTO 800
820 !
830 ! ******************* HOT SOURCE TEMP *******************
840 DISP "Enter temperature (deg K) of noise source";@ INPUT TNS
850 PR$="Temp. of noise source = "&VAL$ (TNS)&" deg K"&CS$ @ GOTO 910
860 !
870 ! ********************** ENTER ENR ********************
880 DISP "Enter excess noise ratio (dB)";@ INPUT ENR
890 IF PN#1 THEN PRINT "Excess noise ratio = ";ENR;" dB"
900 TNS=290*(FNIPDB(ENR)+1) @ GOTO 850
910 GOSUB 1110 @ DISP "Select Noise figure meas. method" @ DISP
920 DISP "1  3dB Method"
930 DISP "2  Y-factor Method"
940 DISP "Enter #";@ INPUT R@ IF R=2 THEN 1040
950 IF R<1 OR R>2 THEN DISP "1 or 2?" @ BEEP @ GOTO 940
960 !
970 ! ***************** ENTER 3-DB METHOD RATIO ****************
980 DISP "Enter attenuator power ratio(<1)";Q$;@ INPUT ALPH
990 IF ALPH=0 THEN 260
1000 TE=TNS*ALPH-TC @ NFDB=FNPDB(1+TE/290)
1010 PR$="POWER RATIO="&VAL$ (ALPH)&",    EFF. NOISE TEMP ="&VAL$ (TE)
1020 PR$=PR$&"deg K, Noise Figure="&VAL$ (NFDB)&"dB" @ GOSUB 1100 @ GOTO 970
1030 !
```

```
1040 ! ***************    ENTER Y-FACTOR    ************************
1050 DISP "Enter Y-factor (>1)";Q$;@ INPUT Y@ IF Y=0 THEN 260
1060 TE=(TNS-Y*TC)/(Y-1) @ NFDB=FNPDB(1+TE/290)
1070 PR$="Y-factor="&VAL$ (Y)&",    Te= "&VAL$ (TE)&"deg K,   NOISE FIGURE="
1080 PR$=PR$&VAL$ (NFDB)&"dB" @ GOSUB 1100 @ GOTO 1050
1090 !
1100 ! *******************    DISP/PRINT    **************************
1110 DISP PR$ @ IF PN#1 THEN PRINT PR$
1120 RETURN
1130 !
1140 ! ***********   INITIALIZATION &  PRINTER SELECTION   ***************
1150 PAGESIZE 24 @ CLEAR @ DEG @ PRINTER IS 1
1160 DISP TAB (10);T$ @ DISP
1170 DISP "Select Printer:" @ DISP
1180 DISP "1   CRT"
1190 DISP "2   HP-82905B--HP-IB"
1200 DISP "3   HP-82905B--Centronics"
1210 DISP "4   MX-100"
1220 DISP "Enter #";@ INPUT PN
1230 IF PN<1 OR PN>4 THEN DISP "Use # 1-4 " @ BEEP @ GOTO 1220
1235 IF PN#1 THEN DISP "Make sure Top-Of-Form is set properly, Enter
date(no',')";@ INPUT DA$
1240 ON PN GOTO 1330,1260,1270,1290
1250 !
1260 PRINTER IS 301 @ GOTO 1280
1270 PRINTER IS 701
1280 PRINT CHR$ (27)&"&11L" @ GOTO 1320
1290 PRINTER IS 701 @ PRINT CHR$ (27);"N";CHR$ (8)
1320 PRINT TAB (15);T$ @ PRINT TAB (25);DA$
1330 RETURN
```

Program AUTOCOR

```
10 ! Pgm AUTOCOR , Rev 2/05/84, written by W. Skillman
20 ! Translated from Prog. 6-2, Radar Calcs.
30 !
40 ! Improvements vs. Radar Calc...
50 !    Correlation Coefficient
60 !    Skip autocorrelation calc if data is straight line
70 !
80 ! Important variables:
90 !    b.....X-intercept of trendline (regression line)
100 !    m.....slope of trendline
110 !    NPTS..no. of data points
120 !    r.....correlation coef.
130 !    SIG...std. dev.
140 !    SUMIX.sum of i*X(i)
150 !    TFLAG.=0 if trendline not removed, =1 if removed
160 !    X,XX..data arrays, original, trendline removed
170 !    XB2...square of mean
180 !    XBAR..mean
190 !    XSUM..sum of X(i)
200 !    XSUM2.sum of squares of X(i)
210 !
220 DIM X(1000),XX(1000),r(1000),T$[25],PR$[70]
230 !
240 DEF FNRD(X) = INT (X*1000+.5)/1000 ! Round to 3 decimals
250 T$="** AUTOCORRELATION **"
260 !
270 GOSUB 1020 ! init, printer selection
280 XSUM,XSUM2,SUMIX,TFLAG=0 @ I=1 @ PR$="" @ GOSUB 990
290 DISP "Enter data, X(";I;") (C to correct last entry, Q to quit)";
300 INPUT X$@ X$=UPC$ (X$) @ IF X$="C" THEN GOSUB 810 @ GOTO 290
310 IF X$="Q" THEN NPTS=I-1 @ GOTO 330
320 X,X(I)=VAL (X$) @ GOSUB 840 @ I=I+1 @ IF I<1000 THEN 290
330 DISP "List data for check";@ INPUT R$@ R$=UPC$ (R$)
340 IF R$="N" THEN 370
350 IF R$#"Y" THEN DISP "Y or N?" @ BEEP @ GOTO 330
360 FOR I=1 TO NPTS @ DISP I,X(I) @ NEXT I
370 DISP "Any corrections";@ INPUT R$@ R$=UPC$ (R$)
380 IF R$="N" THEN 420
390 IF R$#"Y" THEN DISP "Y or N?" @ BEEP @ GOTO 370
400 DISP "Enter index of data and correct data (0,0 to quit)";@ INPUT I,XX
410 IF I THEN I=I+1 @ GOSUB 810 @ X,X(I)=XX @ GOSUB 840 @ GOTO 400
420 PR$="  Statistics of original data:"
430 GOSUB 990 @ GOSUB 950 ! Mean and standard deviation
440 m=12*(SUMIX/NPTS-(NPTS+1)*XBAR/2)/(NPTS^2-1) ! Slope of trend line
450 r=(SUMIX-(NPTS+1)*XSUM/2)/SQR ((NPTS^2-1)/12*(NPTS*XSUM2-XSUM^2))
460 b=XBAR-m*(NPTS+1)/2
470 rr=FNRD(r) @ br=FNRD(b) @ mr=FNRD(m)
480 PR$="Trendline data: slope= "&VAL$ (mr)&", X(0)= "&VAL$ (br)
490 PR$=PR$&", correl. coef. = "&VAL$ (rr) @ GOSUB 990
500 IF ABS (r)#1 THEN 520
```

```
510 PR$="Data is a straight line, autocorrelation=1" @ GOSUB 990 @ GOTO 790
520 DX=m*(NPTS-1)/SIG ! trend removal indicator
530 DISP "dX/sig=";DX;", do you wish to remove the trend";@ INPUT R$
540 R$=UPC$ (R$) @ IF R$="N" THEN 560
550 IF R$="Y" THEN GOSUB 870 ELSE DISP "Y or N?" @ BEEP @ GOTO 530
560 r(0)=1 @ FOR K=1 TO NPTS-1 @ SUMXX=0 ! autocorrelation function
570   FOR I=1 TO NPTS-K
580     IF TFLAG THEN P=XX(I)*XX(I+K) ELSE P=X(I)*X(I+K)
590     SUMXX=SUMXX+P
600   NEXT I @ r(K)=(SUMXX/(NPTS-K)-XB2)/(XSUM2/NPTS-XB2)
610 NEXT K @ r(K)=0 @ IF TFLAG=0 THEN 720
620 PR$="  Index        Orig Data        Data-trend        Autocorrelation"
630 GOSUB 990
640 PR$="   0                                          1" @ GOSUB 990
650 FOR I=1 TO NPTS
660   DISP USING FOR2 ; I,X(I),XX(I),r(I)
670   IF PN#1 THEN PRINT USING FOR2 ; I,X(I),XX(I),r(I)
680   FOR2: IMAGE X,DDD,6X,DDD.DDD, 8X,DDD.3D,9X,3D.3D
690 NEXT I @ GOTO 790
710 !
720 PR$="  Index        Orig Data        Autocorrelation" @ GOSUB 990
730 PR$="   0            -                1" @ GOSUB 990
740 FOR I=1 TO NPTS
750   IF PN#1 THEN PRINT USING FOR1 ; I,X(I),r(I)
760   DISP USING FOR1 ; I,X(I),r(I)
770   FOR1: IMAGE X,3D,6X,3D.3D,8X,3D.3D,9X,3D.3D
780 NEXT I
790 DISP "New data";@ INPUT R$@ IF R$="Y" THEN 280 ELSE CHAIN "Autost"
800 !
810 ! **************  SUB TO DELETE DATA  SUMS  *******************
820 I=I-1 @ X=X(I) @ XSUM=XSUM-X @ XSUM2=XSUM2-X*X @ SUMIX=SUMIX-I*X @ RETURN
830 !
840 ! **************  SUB TO ENTER DATA SUMS  *******************
850 XSUM=XSUM+X @ XSUM2=XSUM2+X*X @ SUMIX=SUMIX+I*X @ RETURN
860 !
870 ! *************  SUB TO REMOVE TRENDLINE  **********************
880 XSUM,XSUM2,SUMIX=0 @ TFLAG=1
890 FOR I=1 TO NPTS
900  X,XX(I)=X(I)-m*I-b @ GOSUB 840
910 NEXT I @ PR$="  Trendline removed from data, new statistics are:"
920 GOSUB 990
930 !
940 ! *********  SUB TO CALC MEAN AND STD DEV  **************
950 XBAR=XSUM/NPTS @ XB2=XBAR^2 @ SIG=SQR ((XSUM2-NPTS*XB2)/(NPTS-1))
960 PR$="mean="&VAL$ (XBAR)&",  std. dev. ="&VAL$ (SIG)
970 !
980 ! *************  DISP/PRINT  *******************
990 DISP PR$ @ IF PN#1 THEN PRINT PR$
1000 RETURN
1010 !
```

```
1020 ! ***********  INITIALIZATION &  PRINTER SELECTION  ***************
1030 PAGESIZE 24 @ CLEAR @ DEG @ PRINTER IS 1
1040 DISP TAB (10);T$ @ DISP
1050 DISP "Select Printer:" @ DISP
1060 DISP "1   CRT"
1070 DISP "2   HP-82905B--HP-IB"
1080 DISP "3   HP-82905B--Centronics"
1090 DISP "4   MX-100"
1100 DISP "Enter #";@ INPUT PN
1110 IF PN<1 OR PN>4 THEN DISP "Use # 1-4 " @ BEEP @ GOTO 1100
1115 IF PN#1 THEN DISP "Make sure Top-Of-Form is set properly, Enter
date(no',')";@ INPUT DA$
1120 ON PN GOTO 1210,1140,1150,1170
1130 !
1140 PRINTER IS 301 @ GOTO 1160
1150 PRINTER IS 701
1160 PRINT CHR$ (27)&"&11L" @ GOTO 1200
1170 PRINTER IS 701 @ PRINT CHR$ (27);"N";CHR$ (8)
1200 PRINT TAB (15);T$ @ PRINT TAB (25);DA$
1210 DISP @ DISP "Answer questions with 'Y' or 'N', [END LINE]" @ DISP
1220 RETURN
```

Program AMBFUN

```
10 ! Pgm AMBFUN, Rev 2/27/84
20 ! Ambiguity function for phase-coded CW (periodic) or
30 !     pulsed (aperiodic) signal
40 ! Translated  from Pgm 6-3 of Radar Calc...
50 !
60 ! Improvements vs. Radar Calc...
70 !    Gaussian phase errors optional on xmt
80 !    Binary Phase Shifter option (Frank, L-K)
90 !    Plot on plotter or CRT, dump to printer
100 !   Lewis-Kretschmer P2,P3, P4 codes
110 !
120 ! Important variables: (all angles in degrees)
130 !    APFLAG : 0=periodic function, 1=aperiodic
140 !    FT= Doppler freq/PRF
150 !    NB = no. of bits, Binary phase shifter
160 !    NF = Frank code # = sqr(PCR)
170 !    NLK = L-K code # = sqr(PCR)
180 !    PA = Plotter address
190 !    PCR = Pulse Comp. Ratio or bits in code
195 !    PFLAG = 1, Print code, =0, don't
200 !    PH = Rx phase code
210 !    PHRMS = rms random error
220 !    PHRN = Tx phase code (random errors and/or bin p.s. optional)
230 !    PN = Printer Code
240 !
250 DIM T$[30],S$[40],PP$[40]
260 DIM PH(500),PHRN(500),FF(500),HDR$[60],PR$[60],PE$[35]
270 T$="**AMBIGUITY FUNCTION**" @ YL$="Response (dB)"
280 X$="Lag (bits)" @ YN$="Y or N?" @ LK$=" Lewis-Kretschmer "
285 PE$="Rms transmit phase error (deg)"
290 !
300 !
310 DEF FNPDB(X) = 10*LGT (MAX (EPS ,ABS (X))) ! Power in dB,HP-86/87
320 DEF FNRNGAUS(XAVG,rms) = XAVG+rms*SQR (-(2*LOG (MAX (EPS ,RND ))))*COS
(360*RND ) ! Random Gaussian variable
330 !
340 GOSUB 2310 ! init & select Printer
350 GOSUB 2530 ! Select Plotter, load binary pgm GDUMP
360 !
370 DISP "Enter code length, PCR";@ INPUT PCR
380 PR$="Code length="&VAL$ (PCR) @ GOSUB 920
390 DISP "Select code to load"
400 DISP "Automatically loaded codes:"
410 FRAC=FP (SQR (PCR)) @ IF FRAC THEN RM=4 @ GOTO 450 ELSE RM=1
420 DISP "  1 Frank"
430 DISP "   2";LK$;"P1"
440 DISP "   3";LK$;"P2"
450 DISP "   4";LK$;"P3"
460 DISP "   5";LK$;"P4"
470 DISP "Manually loaded codes:"
480 DISP "  6  Binary"
```

```
490 DISP "  7  Quadriphase"
500 DISP "8  New Code Length"
510 DISP @ DISP "Enter #";@ INPUT R@ IF R=8 THEN 370
520 IF R<RM OR R>7 THEN DISP "Use # ";RM;"-8" @ BEEP @ GOTO 510
530 IF R=1 THEN S$,PR$="Frank Code" @ GOTO 560
540 IF R<6 THEN S$,PR$=LK$&"Code P"&VAL$ (R-1) @ GOTO 560
550 DISP "Enter name of code";@ INPUT S$
560 PFLAG=0 @ IF PN=1 THEN 600 ELSE PRINT S$
570 DISP "Print code";@ INPUT R$@ R$=UPC$ (R$)
580 IF R$="Y" THEN PFLAG=1 @ GOTO 600
590 IF R$#"N" THEN DISP YN$ @ BEEP @ GOTO 570
600 ON R GOSUB 1180,1790,1900,1250,1130,2070,2080 ! load code
610 GOSUB 1640 ! Add random and/or binary p.s.errors to tx phase
620 DISP @ DISP "Select type of function"
630 DISP "1  Periodic"
640 DISP "2  Aperiodic"
650 DISP "Enter #";@ INPUT R@ APFLAG=R-1 @ PF=0
660 IF R<1 OR R>2 THEN DISP "Use 1 or 2" @ BEEP @ GOTO 650
670 IF PN=1 THEN 700
680 IF R=1 THEN PR$="P" ELSE PR$="Ap"
690 PRINT PR$;"eriodic function"
700 GOSUB 950 ! Ambiguity function, single Doppler
710 DISP "Select Plot Mode:" @ DISP
712 DISP " 1 No Plot" @ DISP " 2 New Plot"
714 K=2 @ IF PF>1 THEN DISP " 3 Overlay Plot" @ DISP @ K=3
716 DISP "Enter #";@ INPUT PF@ IF PF=1 THEN 780
720 IF PF<1 OR PF>K THEN DISP "Use # 1-";K @ BEEP @ GOTO 716
730 XL=KMIN @ XR=PCR-1 @ AP,YT=0 @ YB=-60 @ LT,DX=1 @ DELY=2
735 ON PF GOSUB 930,1370,1350
740 IF PA=1 THEN DISP "Dump to printer";@ INPUT R$ ELSE 780
750 R$=UPC$ (R$) @ IF R$="N" THEN 780
760 IF R$#"Y" THEN DISP YN$ @ BEEP @ GOTO 740
765 IF PN=1 THEN GOSUB 2350 @ GOTO 765
770 IF PN=4 THEN DUMP GRAPHICS 0,0,0,-1 ELSE DUMP GRAPHICS
780 PR$="" @ GOSUB 920 @ DISP "Main Menu: Select Next Operation"
790 DISP @ DISP "1  New Code length"
800 DISP "2  New Code"
810 DISP "3  Select Aperiodic/periodic"
820 DISP "4  New random errors"
830 DISP "5  More frequencies"
840 DISP "6  Disc Menu"
850 DISP "Enter #";@ INPUT R@ IF R=6 THEN CHAIN "Autost"
860 IF R<1 OR R>6 THEN DISP "Use # 1-6" @ BEEP @ GOTO 850
870 ON R GOTO 370,390,620,610,700
900 !
910 ! ***************** print & display ********************
920 DISP PR$ @ IF PN#1 THEN PRINT PR$
930 RETURN
940 !
950 ! *************** CALC AMBIGUITY FUNCTION ****************
960 DISP "Enter Doppler frequency, fT";@ INPUT FT@ DEG
970 IF PN#1 THEN PRINT "Doppler frequency=";FT
980 DFPH=360*FT/PCR @ KMIN,SUMG=0 @ IMIN=1 @ IMAX=PCR
990 PR$="  K                    G(K)(dB)        TIME(SEC)"
```

```
1000 GOSUB 920 @ PHRN(0)=PHRN(PCR)
1010 IF APFLAG THEN KMIN=1-PCR
1020 FOR K=KMIN TO PCR-1 @ SUMI,SUMQ=0 @ T1=TIME
1030   IF APFLAG THEN IMIN=MAX (1,K+1) @ IMAX=MIN (PCR+K,PCR)
1040   FOR I=IMIN TO IMAX
1050     I2=(I-K) MOD PCR
1060     DPH=PH(I)-PHRN(I2)+I*DFPH
1070     SUMI=SUMI+COS (DPH) @ SUMQ=SUMQ+SIN (DPH)
1080   NEXT I @ X=(SUMI*SUMI+SUMQ*SUMQ)/PCR/PCR
1090   FF(K+1-KMIN)=FNPDB(X)
1100   SUMG=SUMG+X @ PR$=VAL$ (K)&"              "&VAL$ (FF(K+1-KMIN))&"
"&VAL$ (TIME -T1) @ GOSUB 920
1110 NEXT K @ PR$="Sum of clutter power="&VAL$ (SUMG) @ GOTO 920
1120 !
1130 ! ************  SUB TO LOAD LEWIS-KRET P4 CODE  ************
1140 FOR K=1 TO PCR
1150   PH(K),PHRN(K)=180*(K-1)*((K-1)/PCR-1) MOD 360 @ GOSUB 2030
1160 NEXT K @ RETURN
1170 !
1180  ! ************   STORE FRANK CODE  ********************
1190 NF=SQR (PCR)
1200 FOR K=1 TO PCR @ X=(K-1)/NF
1220   PH(K),PHRN(K)=360*INT (X)*FP (X) MOD 360 @ GOSUB 2030
1230 NEXT K @ RETURN
1240 !
1250 ! **********   SUB TO LOAD LEWIS-KRET P3 CODE   **********
1260 KK=180/PCR
1270 FOR K=1 TO PCR
1280   PH(K),PHRN(K)=KK*(K-1)*(K-1) MOD 360 @ GOSUB 2030
1290 NEXT K @ RETURN
1300 !
1310 ! ******************  PLOT DATA  *************************
1320 ! INPUTS : XL=MIN X, XR=MAX X, YT=MAX Y, YB=MIN Y
1350 AP=1 @ DISP "Enter LINE TYPE # 1-8 " @ INPUT LT@ GOTO 1400
1370 IF PA=1 THEN LIMIT 0,125,0,75 @ X2=155 @ X3=85 @ GRAPH @ GCLEAR @ FRAME
ELSE LIMIT 12,252,18,192 @ X2=134 @ X3=73
1380 !
1390 IF AP=0 THEN GOSUB 1540
1400 LOCATE 18,X2,10,84 @ SETGU
1410 IF AP=1 THEN 1470
1420 MOVE X3,89 @ CSIZE 5 @ LDIR 0 @ LORG 4 @ LABEL S$
1430 MOVE 18,86 @ CSIZE 4 @ LORG 2 @ LABEL "Tx errors=",PHRMS," deg rms"
1440 MOVE X2,86 @ LORG 8 @ LABEL "Doppler freq.*T=",FT
1450 MOVE 18,5 @ LORG 3 @ LABEL "Code length=",PCR
1460 MOVE X2,5 @ LORG 9 @ LABEL DA$
1470 SCALE XL,XR,YB,YT
1480 MOVE XL,FF(1) @ I=1 @ LINE TYPE LT
1490 FOR X=XL+DX TO XR STEP DX
1500   I=I+1 @ DRAW X,FF(I)
1510 NEXT X @ IF PA=1 THEN ALPHA @ RETURN ELSE PEN 0 @ MOVE XR,YT @ RETURN
1520 !
```

```
1530 ! ******************** BLANK GRAPH ********************
1540 LOCATE 18,X2,10,84 @ FRAME
1550 MOVE 4,50 @ LORG 4 @ LDIR 90 @ CSIZE 5,.7,30 @ LABEL YL$
1560 MOVE X3,94 @ CSIZE 8 @ LDIR 0 @ LABEL T$
1570 MOVE X3,5 @ LORG 6 @ CSIZE 5 @ LABEL X$
1580 SCALE XL,XR,YB,YT @ CSIZE 4
1590 X=ABS (XL-XR) @ DELX=.2
1600 IF X>10 THEN DELX=.4
1610 IF X>25 THEN DELX=INT (X/25)
1620 FXD 0,0 @ LGRID -DELX,DELY,0,YB,5,5 @ RETURN
1630 !
1640 ! ************* RANDOMIZE PHASE OF TRANSMIT ************
1650 PR$="" @ GOSUB 920 @ IF R<6 THEN GOSUB 2200
1660 DISP "Enter ";PE$ @ INPUT PHRMS
1670 IF PN#1 THEN PRINT PE$;"=";PHRMS
1680 IF PHRMS=0 THEN RETURN
1700 DISP "Enter random # seed (0 to randomize seed)";@ INPUT RS
1710 IF RS THEN RANDOMIZE RS ELSE RANDOMIZE
1730 PR$="Modified Phases are:" @ GOSUB 920
1740 FOR I=1 TO PCR
1750   X,PHRN(I)=FNRNGAUS(PH(I),PHRMS) @ DISP X;@ IF PFLAG THEN PRINT X;
1770 NEXT I @ RETURN
1780 !
1790 ! ************ SUB TO LOAD L-K P1 CODE ***************
1800 NLK=SQR (PCR)
1810 IF FP (NLK) THEN DISP "Use square for L-K P1 code" @ BEEP @ GOTO 370
1820 KK=-(180/NLK) @ K=1
1830 FOR J=1 TO NLK @ J1=(J-1)*NLK-1 @ J2=NLK-2*J+1
1840   FOR I=1 TO NLK
1850     PHRN(K),PH(K)=KK*J2*(J1+I) MOD 360 @ GOSUB 2030 @ K=K+1
1870   NEXT I
1880 NEXT J @ RETURN
1890 !
1900 ! *************** SUB TO LOAD L-K P2 CODE ******************
1910 NLK=SQR (PCR)
1920 IF FP (NLK) THEN DISP "Use square for L-K P2 code" @ BEEP @ GOTO 370
1930 IF FP (NLK/2) THEN DISP "Warning!! High Sidelobes ahead (NLK odd)"
1940 N1=1-1/NLK @ P=90 @ K=1
1950 FOR J=1 TO NLK @ J2=NLK-2*J+1
1960   FOR I=1 TO NLK
1970     PHRN(K),PH(K)=P*(N1-2*(I-1)/NLK)*J2 MOD 360 @ GOSUB 2030 @ K=K+1
1990   NEXT I
2000 NEXT J @ RETURN
2010 !
2020 ! ***************** DISP/PRINT ****************
2030 DISP PH(K);@ IF PFLAG=1 THEN PRINT PH(K);
2040 RETURN
2050 !
```

```
2060 ! **************** MANUAL ENTRY BI-OR QUADRIPHASE CODE ********
2070 BQFLAG=0 @ PP$="Enter 0 or 1 (0 or 180 deg)" @ KPH=180 @ GOTO 2090
2080 BQFLAG=1 @ PP$="Enter 0,1,2,3 (0,90,180,270 deg)" @ KPH=90
2090 IF PFLAG THEN PRINT "Index          Phase (deg)"
2100 FOR I=1 TO PCR
2110   DISP PP$;@ INPUT R
2120   IF R=0 OR R=1 THEN 2160
2130   IF BQFLAG AND (R=2 OR R=3) THEN 2160
2150   DISP "Bad entry!" @ BEEP @ GOTO 2110
2160   PH(I),PHRN(I)=KPH*R
2170   IF PFLAG THEN PRINT I,PH(I)
2180 NEXT I @ IF PFLAG THEN PRINT
2185 RETURN
2190 !
2200 ! ****************** BINARY PHASE SHIFTER ****************
2210 DISP "Binary phase shifter";@ INPUT R$@ R$=UPC$ (R$)
2220 IF R$="N" THEN RETURN
2230 IF R$#"Y" THEN DISP YN$ @ BEEP @ GOTO 2210
2240 DISP "Enter # of bits ,NB, Binary Phaser (LSB(deg)=360/2^NB)";@ INPUT NB
2250 DPH=360/2^NB ! LSB of Binary Phaser
2260 IF PN#1 THEN PRINT NB;" Bit Binary Phase Shifter, LSB=";DPH;" deg"
2270 FOR I=1 TO PCR
2280   PHRN(I)=DPH*INT (PH(I)/DPH+.5)
2290 NEXT I @ RETURN
2300 !
2310 ! *********** INITIALIZATION &  PRINTER SELECTION **************
2320 PAGESIZE 24 @ CLEAR @ DEG @ PRINTER IS 1
2330 DISP TAB (10);T$ @ DISP
2340 DISP "Enter today's date (any comma-less format)";@ INPUT DA$
2350 DISP "Select Printer:" @ DISP
2360 DISP "1  CRT"
2370 DISP "2  HP-82905B--HP-IB"
2380 DISP "3  HP-82905B--Centronics"
2390 DISP "4  MX-100"
2400 DISP "Enter #";@ INPUT PN
2410 IF PN<1 OR PN>4 THEN DISP "Use # 1-4 " @ BEEP @ GOTO 2400
2415 IF PN#1 THEN DISP "Make sure Top-Of-Form is set properly, Press [END
LINE]";@ INPUT PR$
2420 ON PN GOTO 2500,2440,2450,2470
2430 !
2440 PRINTER IS 301 @ GOTO 2460
2450 PRINTER IS 701
2460 PRINT CHR$ (27)&"&l1L" @ GOTO 2490
2470 PRINTER IS 701 @ PRINT CHR$ (27);"N";CHR$ (8)
2490 PRINT TAB (15);T$ @ PRINT TAB (25);DA$
2500 DISP @ DISP "Answer questions with 'Y' or 'N', [END LINE]" @ DISP
2510 RETURN
2520 !
2530 ! ********** Sub to Select plotter, load binary GDUMP **************
2540 DISP "Enter Plotter Address (CRT=1)";@ INPUT PA
2550 IF PA<1 THEN DISP "Use # >=1" @ BEEP @ GOTO 2540
2560 ON ERROR GOTO 2570 @ GDUMP @ GOTO 2620 ! Error if GDUMP not
resident,          displays "Copyright Hewlett-Packard 203.09", if resident
2570 ON ERROR GOTO 2600 @ LOADBIN "GDUMP" @ GOTO 2560 ! several errors possible
2580 ! ERRN=67 indicates GDUMP not on disc
```

```
2590 ! ERRN=130      "        bad disc, no disc, door open, etc.
2600 IF ERRN =67 OR ERRN =130 THEN DISP "Insert disc with binary GDUMP, press
[CONT]" @ PAUSE @ GOTO 2570
2610 DISP "Unknown error,ERRN=";ERRN ;" try again" @ GOTO 2540
2620 OFF ERROR @ PLOTTER IS PA @ RETURN
```

Program CRT

```
10 ! Pgm CRT, rev 2/05/84, written by W. Skillman
20 ! Translated from Program 6-4, Radar Calc...
30 !
40 ! Improvements vs Radar Calc:
50 !     2<= no. of PRFs <=10
60 !     Primality of PRF divisors checked (must be relatively prime for CRT)
70 !     Keyboard calc of Greatest Common Divisor (GCD)
75 !     Validity check on entered ranges
80 !
90 ! Important variables:
100 !    C...CRT factors
110 !    M...PRF divisors
120 !    MM..Modulo factor for CRT
130 !    NP..# of PRFs
140 !    PN..Printer code
150 !    R...Range
160 !
170 DIM PR$[75]
180 !
190 T$="*** CRT ***" @ GOSUB 1020 ! Init, Printer selection
200 PR$="Ranges are in units of range gate or PRF clock spacing" @ GOSUB 500
210 PR$="" @ GOSUB 490 @ DISP "Select Operation" @ DISP
220 DISP "1  PRFs/Ranges"
230 DISP "2  Greatest Common Divisor"
240 DISP "Enter #";@ INPUT R@ IF R=2 THEN 1230
250 IF R#1 THEN DISP "Use 1 or 2" @ BEEP @ GOTO 240
260 GOSUB 960 ! Select # of PRFs
270 GOSUB 640 ! enter PRF divisors, calc Cs for Chinese Rem. Theorem
280 !
290 PR$="" @ GOSUB 490 @ DISP "Main Menu: Select next operation" @ DISP
300 DISP "1  Range from ambiguous ranges"
310 DISP "2  Ambiguous ranges from Range"
320 DISP "3  New PRF divisors"
330 DISP "4  New # of PRFs"
340 DISP "5  Greatest Common Divisor" @ DISP "6  Disc Menu"
350 DISP @ DISP "Enter #";@ INPUT R@ IF R=6 THEN CHAIN "Autost"
360 IF R<1 OR R>6 THEN DISP "Use # 1-6" @ BEEP @ GOTO 350
370 ON R GOTO 400,540,270,260,1240
380 !
390 ! ********** Range from ambig ranges ************************
400 DISP "Enter ambiguous ranges singly ([EL] to Quit) " @ SSUM=0 @ PR$=""
410 FOR I=1 TO NP
420 DISP "Range R";Q$(I);" (<";M(I);")";@ INPUT R$
430 IF R$="" THEN 290 ELSE R=VAL (R$)
440 IF R<0 OR R>M(I)-1 THEN DISP "Use 0<=Range<";M(I) @ BEEP @ GOTO 420
450 PR$=PR$&",  R"&Q$(I)&R$ @ SSUM=SSUM+R*C(I)
460 NEXT I @ R=SSUM MOD MM
470 PR$=PR$&"  Range="&VAL$ (R) @ GOSUB 500 @ GOTO 400
480 !
```

```
490 ! ******************* DISPLAY/PRINT  *******************
500 DISP PR$ @ IF PN#1 THEN PRINT PR$
510 RETURN
520 !
530 ! ********* SUB TO CALC AMB RANGES FROM UNAMBIGUOUS RANGE  ******
540 DISP "Enter Unambiguous Range ([EL] to Quit)";@ INPUT R$
550 IF R$="" THEN 290 ELSE R=VAL (R$)
570 IF R>MM-1 OR R<0 THEN DISP "Use 0<Range<";MM @ BEEP @ GOTO 540
580 PR$="Range="&R$
590 FOR I=1 TO NP
600    RA=R MOD M(I) @ PR$=PR$&",   R"&Q$(I)&VAL$ (RA)
610 NEXT I @ GOSUB 500 @ GOTO 540
620 !
630 ! ************* Sub for PRF divisor entry  **********
640 PR$="" @ GOSUB 500 @ IF PN#1 THEN PR$="PRF divisors: "
650 MM=1 @ FOR I=1 TO NP
660    DISP "Enter divisor: m(";I;")";@ INPUT M
670    IF FP (M) THEN DISP "Use m an integer" @ BEEP @ GOTO 660
680    IF M<= 1 THEN DISP "Use m>1 " @ BEEP @ GOTO 660
690    IF PN#1 THEN PR$=PR$&"  m"&Q$(I)&VAL$ (M)
700    M(I)=M @ MM=MM*M
710 NEXT I
720 FOR I=1 TO NP-1
730    FOR J=I+1 TO NP @ F1=M(I) @ F2=M(J) @ GOSUB 1290 @ IF GCD=1 THEN 780
740       BEEP @ PR$="*** Divisors m"&Q$(I)&VAL$ (M(I))&" and m"
750       PR$=PR$&Q$(J)&VAL$ (M(J))&" not relatively prime," @ GOSUB 500
760       PR$=" having a common factor of "&VAL$ (GCD)&", try a new set!"
770       GOSUB 500 @ GOTO 640
780    NEXT J
790 NEXT I @ IF PN#1 THEN PRINT PR$
810 !
820 ! ******************** CALC C'S  ******************
830 FOR I=1 TO NP @ Q$(I)="("&VAL$ (I)&")=" @ MM(I)=M(I) @ NEXT I
840 PR$=" 0<=  Unambig. Range = [" @ FOR I=1 TO NP
850    N=0 @ D,IM1=1/MM(1)
860    FOR J=2 TO NP @ D=D*MM(J) @ NEXT J @ DEN=FP (D)/100000
870    B1=(N+IM1)/DEN @ B1=INT (B1+.5)/100000
880    IF FP (B1) THEN N=N+1 @ GOTO 870
890    C(I)=B1*MM/M(I) @ PR$=PR$&VAL$ (C(I))&"*R("&VAL$ (I)&")"
900    IF I#NP THEN PR$=PR$&" + "
910    TEMP=MM(1) @ FOR J=1 TO NP-1 @ MM(J)=MM(J+1) @ NEXT J @ MM(J)=TEMP
930 NEXT I @ PR$=PR$&"] MOD "&VAL$ (MM)&"  < "&VAL$ (MM) @ GOTO 500
940 !
950 ! ********* SUB TO SELECT # OF PRFS  *****************
960 PR$="" @ GOSUB 500 @ DISP "Enter # of PRFs (2-10) ";@ INPUT NP
970 IF NP<2 OR NP>10 THEN DISP "Use 0<#<10" @ BEEP @ GOTO 960
980 IF FP (NP) THEN DISP "Use integer" @ BEEP @ GOTO 960
990 IF PN#1 THEN PRINT "Number of PRFs = "&VAL$ (NP)
1000 RETURN
1010 !
```

```
1020 ! ******  INITIALIZATION &  PRINTER SELECTION  *******
1030 PAGESIZE 24 @ CLEAR @ PRINTER IS 1
1040 DISP TAB (20);T$ @ DISP
1050 DISP "Select Printer:" @ DISP
1060 DISP "1  CRT"
1070 DISP "2  HP-82905B--HP-IB"
1080 DISP "3  HP-82905B--Centronics"
1090 DISP "4  MX-100"
1100 DISP "Enter #";@ INPUT PN
1115 IF PN#1 THEN DISP "Make sure Top-Of-Form is set properly, Enter
date(no',')";@ INPUT DA$
1120 ON PN GOTO 1210,1140,1150,1170
1130 !
1140 PRINTER IS 301 @ GOTO 1160
1150 PRINTER IS 701
1160 PRINT CHR$ (27)&"&11L" @ GOTO 1200
1170 PRINTER IS 701 @ PRINT CHR$ (27);"N";CHR$ (8)
1200 PRINT TAB (25);T$ @ PRINT TAB (25);DA$
1210 RETURN
1220 !
1230 ! ************* Keyboard calc of GCD  ****************
1240 DISP "Enter integers I,J (0,0 to quit)";@ INPUT I,J
1250 IF I AND J THEN F1=I @ F2=J @ GOSUB 1290 ELSE 290
1260 PR$="GCD of "&VAL$ (I)&", "&VAL$ (J)&" = "&VAL$ (GCD)
1270 GOSUB 500 @ GOTO 1240
1280 !
1290 ! *************** Greatest common divisor ***********
1300 GCD=F1 @ X,R4=F2
1310 TEMP=GCD @ GCD=X @ X=TEMP
1320 X=GCD*FP (X/GCD) @ X=INT (X*100000+.5)/100000
1330 IF X THEN 1310 ELSE RETURN
```

```
460 DISP "1  Digital Filter Analysis(DIGFILT)"
470 DISP "2  Butterworth Digital Filter (DIGFILT)"
480 DISP "3  Chebyshev     ''       ''        '' "
490 DISP "4  FFT (DFT)"
500 DISP "5  DFT   '' "
510 DISP "6  Dolph-Chebyshev Wts/Response (DOLPH)"
520 DISP "7  Paired Echo Anal. (PAIRECHO)"
530 DISP "8  Taylor Wt Coefs (PAIRECHO)"
540 DISP "9  FIR Canc Resp (FIRCANC)"
550 DISP "10 MEM Spectrum (MEM)"
560 DISP "11 Imp Fact-DFT (IMPFACT)"
570 DISP "12 Imp Fact-FIR Clut Canc (FIRCANC)"
580 DISP @ DISP "Enter #";@ INPUT R
590 IF R<1 OR R>12 THEN DISP "Use # 1-12" @ BEEP @ GOTO 580
600 ON R GOTO 620,620,620,630,630,640,650,650,670,680,690,670
610 !
620 CHAIN "DIGFILT"
630 CHAIN "DFT"
640 CHAIN "DOLPH"
650 CHAIN "PAIRECHO"
670 CHAIN "FIRCANC"
680 CHAIN "MEM"
690 CHAIN "IMPFACT"
700 !
710 DISP "1  Noise Fig/Temp (NOISEFIG)"
720 DISP "2  Autocorrelation Fn (AUTOCOR)"
730 DISP "3  Amb Fn: Phase-coded sig. (AMBFUN)"
740 DISP "4  Mult PRF Range Res. (CRT)"
750 DISP @ DISP "Enter #";@ INPUT R
760 IF R<1 OR R>4 THEN DISP "Use # 1-4" @ BEEP @ GOTO 750
770 ON R GOTO 790,800,810,820
780 !
790 CHAIN "NOISEFIG"
800 CHAIN "AUTOCOR"
810 CHAIN "AMBFUN"
820 CHAIN "CRT"
```

Chapter 7 Autost Programs

Program Autost

```
10 ! Pgm Autost, rev. 11/22/83
12 ON ERROR GOTO 14 @ GDUMP @ GOTO 16
14 LOADBIN "GDUMP"
16 OFF ERROR @ PAGESIZE 24 @ CLEAR
20 DISP @ DISP @ DISP @ DISP TAB (23);"*** RADAR CALCULATIONS ***"
30 DISP @ DISP TAB (29);"by W. Skillman" @ DISP
40 DISP TAB (29);"November, 1983" @ DISP @ DISP @ DISP "** Disc Menu **"
50 DISP @ DISP "Select Chapter:" @ DISP
60 DISP "1  Antennas"
70 DISP "2  Propagation"
80 DISP "3  Detection Probability"
90 DISP "4  Signal-to-noise Calcs."
100 DISP "5  Filters & Filtering"
110 DISP "6  Receivers & processing"
120 DISP @ DISP "Enter #";@ INPUT R
130 IF R<1 OR R>6 THEN DISP "Use # 1-6" @ BEEP @ GOTO 120
140 DISP @ DISP "Select Program in Chapter";R;":" @ DISP
150 ON R GOTO 170,300,420,440,460,710
160 !
170 DISP "1  Antenna Parameters (ANTPAR)"
180 DISP "2  Scanned Array Pattern (PATPLOT)"
190 DISP "3  Array Ant. Beam Calcs (ANTBEAM)"
200 DISP "4  Ant Pattern Synthesis (ANTSYN)"
210 DISP @ DISP "Enter #";@ INPUT R
220 IF R<1 OR R>4 THEN DISP "Use # 1-4" @ BEEP @ GOTO 210
230 ON R GOTO 250,260,270,280
240 !
250 CHAIN "ANTPAR"
260 CHAIN "PATPLOT"
270 CHAIN "ANTBEAM"
280 CHAIN "ANTSYN"
290 !
300 DISP "1  Radar to tgt angles & ranges (ANGLES)"
310 DISP "2  Radar RHA chart,exp atmos. (TRACE)"
320 DISP "3  Radar Covg. Diag.(COVER)"
330 DISP "4  Low Alt Tgt Track Error, Monopulse (MONOERR)"
340 DISP @ DISP "Enter #";@ INPUT R
350 IF R<1 OR R>4 THEN DISP "Use # 1-4" @ BEEP @ GOTO 340
360 ON R GOTO 370,380,390,400
370 CHAIN "ANGLES"
380 CHAIN "TRACE"
390 CHAIN "COVER"
400 CHAIN "MONOERR"
410 !
420 CLEAR @ DISP "Loading DETPROB " @ CHAIN "DETPROB"
430 !
440 CLEAR @ DISP "Loading SIG/CLUT" @ CHAIN "SIG/CLUT"
450 !
```

Program Autost 1

```
10  ! Pgm Autostl, rev 2/12/84, written by W. Skillman
20  PAGESIZE 24 @ CLEAR @ FOR X=5 TO 14 @ ON KEY# X GOTO 340 @ NEXT X
30  ON ERROR GOTO 40 @ GDUMP @ GOTO 50
40  LOADBIN "GDUMP"
50  ON ERROR GOTO 60 @ AWRITE 0,0 @ GOTO 80
60  ON ERROR GOTO 70 @ LOADBIN "UTIL/1" @ GOTO 80
70  DISP "Insert Disc with UTIL/1, press CONT, then re-insert Radar Calculations
Disc" @ PAUSE @ GOTO 60
80  OFF ERROR @ CLEAR @ TAKE KEYBOARD @ DIM A$[80],B$(6,12)[65]
90  A$=RPT$ ("*",80) @ AWRITE 0,0,A$
100 FOR I=1 TO 23 @ AWRITE I,0,"*" @ AWRITE I,79,"*" @ NEXT I @ AWRITE 23,0,A$
110 AWRITE 5,22,"***  Radar Calculations  ***"
120 AWRITE 10,27,"by W. Skillman" @ AWRITE 15,27,"February, 1984"
125 AWRITE 19,22,"Copyright Artech House"
130 DATA Antennas,Propagation,Detection Probability (DETPROB)
140 DATA Signal; Clutter; Jamming (SIG/CLUT),Filters & Filtering
150 DATA Receivers & Processing,Antenna Parameters (ANTPAR)
160 DATA Scanned Array Pattern (PATPLOT),Array Antenna Beam Calcs (ANTBEAM)
170 DATA Antenna Pattern Synthesis (ANTSYN),Radar to target angles & ranges
(ANGLES),Radar RHA chart; exponential atmosphere (TRACE)
180 DATA Radar Coverage Diagram (COVER)
190 DATA Monopulse Low Altitude Target Tracking Error (MONOERR)
200 DATA Digital Filter Analysis (DIGFILT),Butterworth Digital Filter (DIGFILT)
210 DATA Chebyshev Digital Filter (DIGFILT),FFT (DFT),DFT (DFT)
220 DATA Dolph-Chebyshev Weights/Response (DOLPH),Paired Echo Analysis
(PAIRECHO),Taylor Weighting Coefficients (PAIRECHO),FIR Canceller Response
(FIRCANC)
230 DATA MEM Spectrum (MEM),Improvement Factor for DFT (IMPFACT)
240 DATA Improvement Factor for Clutter Canceller (FIRCANC),Noise
Figure/Temperature (NOISEFIG)
250 DATA Autocorrelation Function (AUTOCOR),Ambiguity Function: Phase-coded
signal (AMBFUN),Multiple PRF Range Resolution (CRT)
260 J=0 @ NL=6 @ GOSUB A  @ J=1 @ NL=4 @ GOSUB A  @ J=2 @ GOSUB A
270 J=5 @ NL=12 @ GOSUB A  @ J=6 @ NL=4 @ GOSUB A  @ WAIT 4000
280 ON KEY# 1,"   UP" GOTO 510 @ ON KEY# 2," DOWN" GOTO 520
290 ON KEY# 3," SELECT" GOTO 370
300 CLEAR @ A$=A$[1,15] @ AWRITE 3,15,A$ @ AWRITE 5,15,A$
310 AWRITE 4,15,"* Disc Menu  *" @ AWRITE 7,17,"Select Chapter:"
320 ND=6 @ J,FLAG=0 @ P=15 @ M=9 @ ON KEY# 4,"" GOTO 340
330 K=0 @ FOR I=1 TO ND @ GOSUB B  @ NEXT I @ N=0 @ KEY LABEL
340 I=N+1 @ K=1 @ GOSUB B  ! HIGHLIGHT line
350 GOTO 350 ! Idle loop
360 !
370 CLEAR @ IF FLAG OR N=2 OR N=3 THEN 420
380 FLAG,M=7 @ J=N+1 @ ON KEY# 4,"Disc Menu" GOTO 300
390 ND=4 @ IF N=4 THEN ND=12
400 AWRITE 4,17,"Select Program in Chapter "&VAL$ (J)&":  "&B$(0,J) @ GOTO 330
410 !
```

```
420 RELEASE KEYBOARD @ A$=B$(J,I) @ A$=A$[POS (A$,"(")+1,POS (A$,")")-1]
430 AWRITE 10,25,"Loading Program "&A$ @ CHAIN A$
440 !
450 A: FOR I=1 TO NL @ READ B$(J,I)@ NEXT I @ RETURN
460 !
470 K=0 @ FOR I=1 TO ND @ GOSUB B  @ NEXT I @ RETURN
480 !
490 B: AWRITE M+I-1,P,HGL?$ (VAL$ (I)&"   "&TRIM$ (B$(J,I)),K) @ RETURN
500 !
510 L=-1 @ GOTO 530
520 L=1
530 K=0 @ GOSUB B  @ N=(N+L) MOD ND @ GOTO 340
```

```
420 RELEASE KEYBOARD σ A$=B$(J,I) σ A$=A$[POS (A$,"(")+1,POS (A$,")")-1]
430 AWRITE 10,25,"Loading Program "∂A$ σ CHAIN A$
440 :
450 FOR I=1 TO NL σ READ B$(J,I)σ NEXT I σ RETURN
460 :
470 K=0 σ FOR I=1 TO ND σ GOSUB 490 σ NEXT I σ RETURN
480 :
490 AWRITE M+I-1,P,HGLμ$ (VAL$ (I)∂"   "∂TRIM$ (B$(J,I)),K) σ RETURN
500 :
510 L=-1 σ GOTO 530
520 L=1
530 K=0 σ GOSUB 490 σ N=(N+L) MOD ND σ GOTO 340
```

Appendix A

A SUMMARY OF HP-86/87 BASIC

This appendix summarizes the BASIC functions and commands that are used in the programs of this supplement. Because the purpose is to help translate these HP programs to other BASIC dialects, and because of the uniqueness and complexity of the graphic functions, these are omitted from this description.

Arithmetic Operations

+   Addition
−   Subtraction
*   Multiplication
/   Division
∧   Exponentiation
MOD   Modulo (A MOD B = A-B*INT(A/B), or the remainder resulting from normal division)

Predefined Functions

ABS(X)	Absolute value of X
ACS(X)	Arccosine of X, in 1st or 2nd quadrant
ASN(X)	Arcsine of X, in 1st or 4th quadrant
ATN(X)	Arctangent of X, in 1st or 4th quadrant
ATN2(Y,X)	Arctangent of Y/X in proper quadrant
CHR$(X)	Character whose decimal character code is X
COS(X)	Cosine of X
EPS	Smallest Positive machine #, (1E-499)
ERRN	Number of latest error
EXP(X)	e to the X power
FP(X)	Fractional part of X
INF	Largest pos. machine # (9.99999999999E499)
INT(X)	Largest integer<=X (NOT the same as INT on TI-59!)
IP(X)	Integer part of X (same as INT on TI-59, not used here)
LEN(S$)	Length of string
LGT(X)	Log to the base 10 of X, X>0
LOG(X)	Log to the base e of X, X>0 (Natural logarithm)
MAX(X,Y)	if X>Y then X, else Y is returned
MIN(X,Y)	if X<Y then X, else Y is returned
PI	3.14159265359
POS(S$,T$)	Returns index of 1st occurrence of string T$ in string S$, 0 if not found
SGN(X)	Signum function: -1 if X<0, 0 if X=0, 1 if X>0
SIN(X)	Sine of X
SQR(X)	Positive square root of X, X>0
TAB(X)	Skip to column X on display or printer
TAN(X)	Tangent of X
TIME	Time in seconds since turn-on (unless set later)
UPC$(S$)	Converts all lower case characters in S$ to upper case
VAL(S$)	Returns numeric equivalent of the string S$
VAL$(X)	String equivalent of X

BASIC Statements

BEEP	Sounds a tone as specified by parameters
CHAIN	Loads and runs specified program, Binary programs are not scratched
CLEAR	Clears CRT
DATA constant	Data to be read by a READ statement
DEF FN	Defines a function (only single statement functions used in these programs)
DEG	Sets Degree mode
DIM	Dimensions arrays and string lengths
DISP	Outputs item(s) to display
DISP USING	Outputs item(s) to display in format specified by string
FN	Beginning of name of Defined Function variable
FOR...TO...STEP	Performs following code until NEXT... is encountered, loops until STEP past end value
GOSUB XXX	Branches execution to line number XXX, and ending with RETURN, then returns to statement following the GOSUB XXX
GOTO XXX	Branches execution to line number XXX
IF...THEN...ELSE	If expression following IF is true (=1), then executes statement(s) following THEN; if false (=0) then executes statement(s) following ELSE
IMAGE	Identifies Format string for DISP(or PRINT) using
INPUT	Stops execution, displays ?, waits for input(s) followed by [END LINE]
INTEGER	Specifies variables as 5 digit integers
KEY LABEL	Displays labels for Softkeys at bottom of CRT
LOADBIN	Loads Binary program from disc
OFF ERROR	turns off branching on error
OFF KEY #	turns off Softkey branching
ON X GOTO 10,20	Branches to statement #, corresponding to value of X (1 or 2 as shown)
ON KEY #	Branch using Softkey
ON ERROR	Branch on error
PAGESIZE XX	Selects 16 or 24 line display
PAUSE	Stops execution, continue with CONT key
PRINT	Outputs to printer specified by PRINTER IS
PRINTER IS XX	Specifies printer interface
RAD	Sets Radian mode
READ	Reads variable from DATA statement
RESTORE XXX	Sets pointer for READ to line XXX
RETURN	Causes return from GOSUB call
WAIT X	Stops execution for X milliseconds

Appendix B

TRANSLATION OF HP-86/87 BASIC TO IBM PC BASIC

The HP-86/87 programs in this supplement were designed for ease of translation to other BASIC dialects. The programs are modularized and annotated so that the user can understand the program flow sufficiently to modify for his own use. The construction is such that many changes required for compatibility with the IBM PC BASIC, which is a version of Microsoft BASIC, can be accomplished by simple substitutions. Suggested substitutions that can be used in the programs of this supplement are described below.

Direct Substitution

Many operators, functions and keywords can be substituted for on a one-for-one basis, as shown in Table B-1.

Table B-1
Direct Substitutions for Operators, Functions and Keywords

Operator/Funct/Keyword	HP-86/87	IBM PC
string concatenation	&	+
remark delimiter	!	'
multiple statement concatenation	@	:
inequality	# or < >	< >
clear screen	CLEAR	CLS
display on CRT	DISP	PRINT
smallest number (1.E-499)	EPS	0.59 E-38
largest ν (9.99999999999E499)	INF	1.7 E38
pi (3.14159265359)	PI	3.141593*
output to printer	PRINT	LPRINT
tabulate to position N	TAB(N)	SPC(N)
numeric to string conversion	VAL$(X)	STR$(X)
displayed Δ	CHR$(7)	"increment"

*If pi is defined as a variable, i.e., PI = 3.141593, then up to 16 decimal places may be used in "double precision." However, the arithmetic functions ATN, COS,EXP, LOG, SIN, SQR, TAN return values in "single precision," or 7 decimal places.

Defined Function Substitution

Many of these HP-86/87 programs take advantage of the greater number of arithmetic functions on this machine. Translation to IBM can utilize defined functions to substitute for the HP internal functions, as shown in Table B-2. These functions can be used the same as predefined functions, i.e. to convert power, P, to decibels, PDB:

HP-86/87: PDB=10*LGT(P)
IBM PC : PDB=10*FNLGT(P)

Table B-2
Defined Function Substitution

Function	HP-86/87	IBM PC
fractional part	FP (X)	DEF FNFP(X)=SGN(X)*(ABS(X)–INT(ABS(X)))
log base 10	LGT(X)	DEF FNLGT(X)=LOG(X)/LOG(10)

Subroutine Substitution

Several HP BASIC functions can be replaced with a subroutine call. Suggested subroutines are shown in Table B-3. Care must be used to avoid duplication of variables since the subroutine variables are global, not local as are the arguments of the defined function calls. The value of pi is assumed to be assigned to the variable, PI, earlier in the program. The subroutines in the table are in a format to be added at the end of a translated program to avoid conflict with current line numbers.

Table B-3
Subroutine Substitution

Function	HP-86/87	IBM PC
		9000 ' Subroutines for HP conversion
arcsine	ASN (X)	9005 ' Sub to calc arcsine (X)
		9010 IF ABS(X) = 1 THEN ASN = PI/2*SGN(X): RETURN
		9020 ASN = ATN (X/SQR (1 – X*X)): RETURN
		9030 '
arccosine	ACS (X)	9035 ' Sub to calc arccosine(X)
		9040 SX = PI* (1 – SGN(X))/2
		9050 IF X = 0 OR ABS(X) = 1 THEN ACS = SX/2: RETURN
		9060 ACS = ATN(SQR (1 – X*X)/X) + SX: RETURN
		9070 '
arctangent (2 argument)	ATN2(Y,X)	9080 ' Sub to calc arctan (Y/X)
		9090 ' in proper quandrant, AT2 is output
		9100 IF X THEN AT2 = ATN (Y/X) + PI/2*(1 – SGN(X)) ELSE AT2 = PI/2*SGN(Y)
		9110 IF AT2 < 0 THEN AT2 = AT2 + 2*PI
		9115 RETURN
		9120 '
minimum of X,Y	MIN(X,Y)	9130 'Sub to calc min of X,Y
		9140 IF X > Y THEN MIN = Y ELSE MIN = X
		9150 RETURN
		9160
maximum of X,Y	MAX(X,Y)	9170 ' Sub to calc max of X,Y
		9180 IF X > Y THEN MAX = X ELSE MAX = Y
		9190 RETURN

Miscellaneous differences

As noted in the footnote to Table B-1, most arithmetic functions used in the IBM PC return single precision (7 place) results. Therefore, many calculations on the IBM PC will have less accuracy than the TI-59 or HP-86/87 which have 13 and 12 place accuracy, respectively.

Both upper case and lower case letters are allowed in variable names in the HP-86/87, but only upper case in the IBM PC (numbers are allowed in both BASICs, following the first character which must be a letter). Some programs in this supplement use lower case letters for consistency with notation in the TI-59 book, but there should be no duplication of names if they are simply changed to upper case. In these programs, a string variable input, such as "n" or "y" is converted to upper case by the operator UPC\$, for convenience in testing for branching. This conversion statement must be deleted in translation to IBM.

DEGree mode is used in most of these BASIC programs, as in the TI-59 programs, due to the author's predeliction for this mode. However, Microsoft BASIC does not permit this luxury, so all angles used in trigonometric expressions must be converted to Radians. To ease the burden, I suggest defining a variable=180/pi to use to convert the input and output angles in the appropriate direction.

In the HP-86/87 string variables longer than 18 characters must be dimensioned in a DIM statement. (This is in addition to the dimensioning of subscripted string variables) Since the IBM PC uses variable length dynamic allocation, these dimension statements must be deleted. Only the string variables using string dimensions indicated by the square bracket [ ] or the bracketed string dimension in a subscripted string variable (string array) also dimensioned in regular parentheses ( ) should be deleted i.e., DIM

A$(15)[40] must be changed to DIM A$(15) to retain the maximum subscript of 15, but DIM B$[40] should be deleted entirely. There are several ports for printer connection in the HP-86/87 so that the desired printer must be specified by a PRINTER IS XXX statement. The selection of XXX and the sending of appropriate print codes to the printer to skip the perforation are handled by a subroutine resident in all the programs of this book. This subroutine also selects 24 lines in the CRT display and enables DEGree mode. The subroutine should be modified to delete the two HP printer options for the IBM PC.

Generation of screen or HP plotter graphics is relatively simple with the many built-in graphics commands in the HP-86/87. The CRT high-level graphics display can also be "dumped" to a printer using a the "binary" program **GDUMP**. Again, this part of the programs in this book is a subroutine which can be deleted in translation, or an equivalent one substituted based on the graphics capability of the IBM PC or peripheral plotter. Many plotters, including Hewlett-Packard, can be used with relatively simple print commands. Appendix D illustrates the type of plot subroutine that can be used to drive a low-cost plotter such as the Radio Shack CGP-115. Another possibility is the use of a relatively crude line printer plot, an example of which is found in program **PATPLOT.**

HP BASIC provides formatted output by use of the DISP USING and PRINT USING statements. The string variable that establishes the format can be either an IMAGE statement or an imbedded string. The programs int his book use the IMAGE statement method. Conversion to IBM BASIC requires that the IMAGE statement be converted to a string variable and a dollar sign ($) be added to the name of the IMAGE In the corresponding PRINT or LPRINT USING statement. Table B-4 shows a translation of the HP IMAGE statements found in several programs to the corresponding IBM BASIC string variables. The first line of each program is included for easy identification and the IBM statements are shown as remarks below the HP IMAGE statement.

Table B-4
Translation of IMAGE Statements to IBM String Variables

```
10 ! Pgm TRACE,2/05/84 calculates radar range-height, angle chart for
1290 F1: IMAGE 3X,DDD.DDD,3X,3DC3D,5X,MDD.DD,3X,DDD.D,3X,M3DC3D,3X,DD.DD
1291 ! F1$="   ###.###     ######,.    +##.##   ###.#   +######,.   ##.##"
1950 F3: IMAGE 2X,3DC3D,3X,D.DDDE,3X,D.DDDE,3X,D.DDDE,3X, D.DDDE
1951 ! F3$="  ######,.   #.###^^^^   #.###^^^^   #.###^^^^   #.###^^^^"


10 ! Pgm DETPROB, Rev 2/27/84
1890 F1: IMAGE 6X,D.DDDE,3X,D.3D
1891 ! F1$="      #.###^^^^   #.###"
1990 F2: IMAGE 10X,3D.2D,11X,D.3D
1991 ! F2$="          ###.##          #.###"
2120 F3: IMAGE 8X,4D.DD,8X,D.3D
2121 ! F3$="        ####.##        #.###"


6640 P4: IMAGE "Single look detection, RHO=",2D
6641 ! P4$="Single look detection, RHO= ##"
6650 P2: IMAGE D,"-out-of-",D," detection,  ",DD," cell window, RHO=",2D
6651 ! P2$="#-out-of-# detection,  ## cell window, RHO=##"
6690 P5: IMAGE "Pfa=",D.DDE,"  n=",D.3DE,"  N=",5D
6691 ! P5$="Pfa=#.##^^^^,  n=#.###^^^^,  N=#####"
6700 P1: IMAGE "Pfa=",D.DDE,"  PFR=",D.DDE,"  n=",D.3DE,"  N=",5D
6701 ! P1="Pfa=#.##^^^^,  PFR=#.##^^^^,  n=#.###^^^^,  N=####"
6720 P3: IMAGE "Yb=",3D.DD
6721 ! P3$="Yb=###.##"


10 ! Pgm SIG/CLUT 2/28/84  SURFACE AND VOLUME CLUTTER, and S/N,S/J, J/N
5960 F1: IMAGE 5D.3D,10X,4D.3D
5961 ! F1$=#####.###          ####.###"
```

```
10 ! Pgm DFT, rev 2/27/84, Written by W. SKILLMAN
1690 FOR2: IMAGE 3X,"Max Amplitude=",M4D.4D,"dB or =",MD.4DE,"(numeric)"
1691 ! FOR2$="    Max Amplitude= +####.#### dB or = +#.####^^^^(numeric)"
1720 FOR1: IMAGE 2X,D.4D,3X,MD.4DE,2X,M5D.3D,2X,M4D.4D,2(2X,MD.4DE)
1721 ! FOR1$="  #.####    +#.####^^^^   +#####.###   +####.####   +#.####^^^^"
1722 ! FOR1$=FOR1$&"   +#.####^^^^"
1760 FOR3: IMAGE /70("*")
1761 ! FOR3$="*********************************" @ FOR3$=FOR3$&FOR3$


10 ! Pgm AUTOCOR , Rev 2/05/84, written by W. Skillman
680    FOR2: IMAGE X,DDD,6X,DDD.DDD, 8X,DDD.3D,9X,3D.3D
681 ! FOR2$=" ###        ###.###         ###.###          ###.###"
770    FOR1: IMAGE X,3D,6X,3D.3D,8X,3D.3D,9X,3D.3D
771 ! FOR1$=" ###        ###.###         ###.###          ###.###"
```

One handy feature of HP-86/87 BASIC is the ability to assign a value to multiple variables in a single statement i.e., A,B,C=5 assigns the value "5" to the variables A,B, and C. In IBM BASIC this must be written as A=5:B=5:C=5.

Program PAIRECHO

This version of the Program **PAIRECHO,** found in Chapter 4, illustrates a typical translation of an HP-86 BASIC program to the IBM PC BASIC. Two defined functions have been added, FNFP and FNLGT. A short cut is used in FNPDB; that is, the function MAX was replaced by the addition of the smallest machine number, .59E-38, to the argument to avoid error when the argument is zero. The lower case variables r and t have been replaced by the variables RT and TI respectively. The variable PI is assigned the value 3.141593 in line 290. The HP printer selections have been removed from the printer selection subroutine.

```
10 ' Pgm PAIRECHO, rev 11/01/83, Translated to IBM, 12/19/83
20 ' trans from Pgms 5-7, 5-8 of Radar Calc. by W. Skillman
30 '
40 ' Improvements vs Radar Calc.
50 '    Bandwidth ratio, weighting loss calc. on coef. entry
60 '
70 ' Important Variables:
80 '    ADB...time resp (dB)
90 '    B.....unweighted noise bandwidth
100 '   BR....bandwidth ratio
110 '   BT....normalized time
120 '   F(I)..Fourier Coef array, i=1 to N
130 '   FR....norm. Freq.
140 '   GDB...freq resp (dB)
150 '   GSL...mainlobe/sidelobe in dB, Taylor wts
160 '   LW....Wt. loss
170 '   N.....No. of Fourier Coefs.
180 '   RT....mainlobe/sidelobe voltage ratio, Taylor wts
190 '   R4....n-bar, Taylor wts
200 '   SIG...sigma, Taylor
210 '
220 DIM F(99)
230 '
240 ' ****************** Defined Functions ***************
250 DEF FNPDB(P) = 10*FNLGT(P+.59E-38) ' Power ratio to dB
260 DEF FNVDB(V) = FNPDB(V*V) ' Voltage ratio to dB
270 DEF FNSIG(X) = SQR(X*X/(ASQ+(X-.5)*(X-.5))) ' Sub 107, pgm 5-8
```

```
275 DEF FNFP(X) = SGN(X)*(ABS(X)-INT(ABS(X))) ! Fractional part
278 DEF FNLGT(X) = LOG(X)/LOG(10) ! Log base 10
280 '
290 T$="**PAIRED ECHO ANALYSIS**" : PI=3.141593
300 GOSUB 900 ' init, printer sel.
310 GOSUB 430 ' enter Fourier coefs.
320 '
330 ' ********** Main Menu **************
340 PR$="" : GOSUB 560
350 PRINT "Select next operation" : DISP
360 PRINT "1  New coefficients"
370 PRINT "2  Time Response"
380 PRINT "3  Frequency response" : PRINT "4  Disc Menu"
390 PRINT : PRINT "Enter #";: INPUT R: IF R=4 THEN CHAIN "Autost"
400 IF R<1 OR R>4 THEN PRINT "Use # 1-4" : BEEP : GOTO 390
410 ON R GOSUB 430,700,800 : GOTO 340
420 '
430 ' ************* Fourier Coef Entry ******************
440 SUM2F=.5 : SUMF=.5 : PR$="" : GOSUB 560
450 PRINT "Select Coefs." : DISP
460 PRINT "1  User-Defined"
470 PRINT "2  Taylor"
480 PRINT : PRINT "Enter #";: INPUT R
490 IF R<1 OR R>2 THEN PRINT "Use 1 or 2" : BEEP : GOTO 480
500 ON R GOSUB 1130,1230
510 '
520 LW=2*SUM2F : LWDB=FNPDB(LW) : BR=LW/4/SUMF/SUMF
530 PR$="Weighting loss="+STR$(LWDB)+"dB, Noise bw ratio="+STR$(BR)
540 '
550 ' **************** display/print ****************
560 PRINT PR$ : IF PN<>1 THEN LPRINT PR$
570 RETURN
580 '
590 ' ********** SUB TO CALC SINC(t) ***************
600 IF TI=0 THEN SINC=1 : RETURN
610 SINC=0 : IF FNFP(TI) THEN SINC=SIN(PI*TI)/PI/TI
620 RETURN
630 '
640 ' ********** Sub to calc time response  at time BT *******
650 TI=BT : GOSUB 590 : TSUM=SINC : FOR I=1 TO N
660   TI=BT+I : GOSUB 590 : TT=SINC : TI=BT-I : GOSUB 590
670   TSUM=TSUM+F(I)*(TT+SINC)
680 NEXT I : ADB=FNVDB(TSUM) : RETURN
690 '
700 ' *************** Time Response ******************
710 PRINT "Enter norm. start time, increment ";: INPUT BT,BDT
720 IF BDT<= 0 THEN PRINT "Use increment > 0" : BEEP : GOTO 710
730 FOR J=1 TO 10
740   GOSUB 640 ' Time Response, SINC(t)
750   PR$="Time="+STR$(BT)+", Response="+STR$(ADB) : GOSUB 560 : BT=BT+BDT
760 NEXT J
770 PRINT "More Responses";: INPUT R$ : IF R$="Y" THEN 730
780 IF R$="N" THEN RETURN ELSE PRINT "Y or N?" : BEEP : GOTO 770
790 '
800 ' ***************** Frequency Response ***************
810 PRINT "Enter norm. freq. increment";: INPUT DF
```

```
820 IF DF<= 0 THEN PRINT "Use increment > 0" : BEEP : GOTO 810
830 FOR FR=0 TO PI STEP 2*PI*DF : FSUM=.5
840   FOR I=1 TO N
850     FSUM=FSUM+F(I)*COS(I*FR)
860   NEXT I : GDB=FNVDB(FSUM/SUMF) : PR$="Normalized freq.="
870   PR$=PR$+STR$(FR/2/PI)+", Response="+STR$(GDB)+"dB" : GOSUB 560
880 NEXT FR : RETURN
890 '
900 ' ******   INITIALIZATION &  PRINTER SELECTION   *******
910 CLS
920 PRINT SPC(10);T$ : DISP
930 PRINT "Select Printer:" : DISP
940 PRINT "1  CRT"
970 PRINT "2  MX-100"
980 PRINT "Enter #";: INPUT PN
990 IF PN<1 OR PN>2 THEN PRINT "Use 1 or 2" : BEEP : GOTO 980
1000 ON PN GOTO 1090,1050
1010 '
1050 PRINT "Make sure Top-of Form is set properly, enter today's date ";
1060 PRINT "(any comma-less format)"; : INPUT DA$
1070 LPRINT CHR$(27);"N";CHR$(8)
1080 LPRINT SPC(15);T$ : LPRINT SPC(25);DA$
1090 PRINT : PRINT "Answer questions with 'Y' or 'N', [Enter]" : DISP
1100 RETURN
1110 '
1120 ' **********  Enter User-Defined Coefs.  ***************
1130 PRINT "Enter # of Fourier Coefs.";: INPUT N
1140 IF N<1 OR N>99 THEN PRINT "Use < 0 # < 100" : BEEP : GOTO 1130
1150 IF FNFP(N) THEN PRINT "Use integer" : BEEP : GOTO 1130
1160 FOR I=1 TO N
1170   PRINT "Enter F(";I;")";: INPUT FF
1180   SUMF=SUMF+FF : SUM2F=SUM2F+FF*FF : F(I)=FF
1190   IF PN<>1 THEN LPRINT "F(";I;") = ";FF
1200 NEXT I : RETURN
1210 '
1220 ' ************  Enter Taylor Coefficients  *********
1230 GOSUB 1380 ' Enter GSL, Calc. n-bar minimum, enter n-bar
1240 SIG=FNSIG(R4) : PR$="Sigma = "+STR$(SIG) : GOSUB 560 : R2=1
1250 FOR M=1 TO N
1260   M2=M*M : X=.5 : H8=M2/SIG/SIG
1270   FOR J=1 TO N
1280     X=X*(1-H8/(ASQ+(J-.5)*(J-.5)))
1290     IF J<>M THEN X=X/(1-M2/J/J)
1300   NEXT J : X=R2*X : F(M)=X : R2=-R2
1310   SUMF=SUMF+X : SUM2F=SUM2F+X*X
1320   PR$="F"+STR$(M)+"="+STR$(X) : GOSUB 560
1330 NEXT M : X=RT/SQR(2) : GOSUB 1520
1340 X=SIG*SQR(5.0968*(ASQ-X))
1350 PR$="Widening factor= "+STR$(X) : GOTO 560
1360 '
1370 ' ************  SUB B OF PGM 5-8, ENTER GSL IN DB  *************
1380 PRINT "Enter GSL(dB)";: INPUT X: GSL=ABS(X)
1390 SIG=0 : IF PN<>1 THEN LPRINT "GSL =";-GSL;" dB"
1400 X=10^(GSL/20) : RT=X : R4=1
1410 GOSUB 1520 : ASQ=X ' Calc A^2
```

```
1420 T=SIG : R4=R4+1
1430 SIG=FNSIG(R4) : IF SIG>T THEN 1420
1440 R4M=R4-1 ' Exit when sigma decreases, which defines n-bar minimum
1450 PRINT "Enter n-bar>";R4M;: INPUT R4
1460 IF R4>100 THEN PRINT "Use n-bar <= 100 " : BEEP : GOTO 1450
1470 IF R4<R4M THEN PRINT "Use n-bar > ";R4M : BEEP : GOTO 1450
1480 IF PN<>1 THEN LPRINT PR$;",  n-bar ="+STR$(R4)
1490 N=R4-1 : RETURN
1500 '
1510 ' ******************** SUB B' OF PGM 5-8, CALC A^2 *********
1520 X=SQR(X*X-1)+X : X=LOG(X)/PI : X=X*X : RETURN
```

Appendix C

CORRECTIONS TO "RADAR CALCULATIONS USING THE TI-59"

Preface Change Author's address to P.O. Box 1693, Mail Stop 1105

pp, 10-11 The program does not properly evaluate Eq. 1-2.1 when sin (180 u) is zero. Change steps 174-176 to GTO 0278 and add the program steps shown on page 134 to correct this.

p. 14 Eq. 1-3.9, sign after 1 should be +, not -

p. 15 3rd line should read "...Eq. 1-3.8...", the normal beamwidth, BW_ϕ given by Eq. 1-3.10 is degrees; in the second line below Eq. 1-3.12, the program steps should be 300-304; the alternate values referred to in the 4th line are omitted. In the 1st paragraph, when S=0, the beam will be normal if $(\Delta\phi + F)$ MOD 360 = 0, for all wavelengths.

p. 17 In Ex. 1-3.1., in PRESS column, "LOG" should read "log", in COMMENTS column, 180 is degrees. *Note* that the guide wavelength in Ex. 1-3.4 is calculated for the old K value of 1. If the B and B′ steps are interchanged, so that K is entered before guide wavelength is calculated, the correct value of 4.56515739204 is obtained.

p.24 the 3rd line below Eq. 1-4.1 should read: "...illumination with the peak of the beam at $\Theta = 0$ degrees."

p.29 in COMMENTS column, ϕ should be Θ

p.30 in the title to Fig. 1-4.4, there should be a bar over "n"

p.41 in the footnote "R" should read "ΔR". The derivative of the function to be zeroed, $F(\alpha)$, as given by Eq. 2-1.11 is an approximation that is valid when the altitudes are much smaller than the Earth's radius.

The expression can be made exact by multiplying the 2nd term on the RHS by $1/(1+h_t/R_e)$ and the 3rd term by $1/(1+h_r/R_e)$. *See* below to make these corrections to the program.

p.42 the factor $(1+h_r/R_e)$ should multiply the last term in the square root in Eq. 2.1-23. The program is correct, however.

p.43 1964 is the publication date of the original, Prentice-Hall edition and 1976 is the date of the Artech House edition.

p.44 with the changes to the program as indicated below, the 5th and 6th printouts of Ex. 2-1.1 are -6.778192573 and .5791231973.

p.45 inexplicably, I now get 100.1175136 and 41021.88521 for the 3rd and 4th printouts of Ex. 2-1.3, and 40967.02821 for 3rd printout of Ex. 2-1.4. Possibly, a slightly different value was used earlier for Register 34 (Earth's radius in feet).

p.48

These changes make Eq. 2-1.11 exact as discussed above:
1 delete steps 290, 291, 353, 354, 355;
2 change steps 251-253 and 268-270 to SBR 279, delete step 200
3 insert ÷ RCL 12 in front of step 186;
4 insert ÷ RCL 11 in front of step 181

p.49 change steps 377-379 and 530-532 to SBR 238

p.51 delete "*" in line above Eq. 2-2.1, add to right-hand side of Eq. 2-2.1, "= (n-1) x 10^6"

p.52 in Fig. 2-2.1, Θ' is the angle at the upper end of the 1st segment labeled "INCIDENT RAY", and Θ'' is the angle of the 2nd ray segment, labeled "REFRACTED RAY", to the local horizontal

p.53 in Eq. 2-211, h_0 is the radar altitude, or the initial height of the ray.

p.60 in the 3rd line in the lower 3rd, 10000 should be in the VALUE COLUMN, OPPOSITE 10,000 feet in the COMMENTS column

p.65 Change steps 126-128 to GTO 05 41 to output x, y of ground point

p.70 In User Instructions, step 3b, value displayed is the angle from the normal, $\Theta-\Theta_{no}$, enter gain with respect to peak gain.

p.71 Example 2-3.1 assumes no atmospheric loss

p.77 Eqs 2-4.4, 2-4.5 are in radians, Eq. 2-4.6, change + to -

p.78 the 4th line should read "$\alpha_c=\lambda/(4\pi\sigma_h')$", the 5th line should read "$\sigma_h=...$" Also, add definition of β_0=rms surface slope deviation. To be consistent with annotation on listing, define $a=\alpha_1/2\beta_o$, $c = \alpha_2/\beta_o$, then a' =min(a, c) and c' =max(a, c). Also, the definitions are for ρ^2, not ρ (with subscripts sj and sa)

p.80 at step 6 e), delete "RST" from PRESS column

pp.84-87 pages 84 and 86 should be interchanged, as should 85 and 87; on current page 84, step 261, the subscript should be h, not n, steps 279, 327 should be annotated Fd (a'); step 285 should be annotated Fd (c'); and on current pages 84, 85, and RUN in the card form at the top. On current page 87, step 139, 140 should be annotated K, not n

p.94 in Eq. 3-0.7, the last factor should read: "$e^{kx/\bar{x}}$", note that k and K are greater than zero, and Eq. 0.9 should read "$K'(\bar{X})^2/\sigma^2(X)$"

p.95 note that ρ is defined greater than 1, but is not tested in program 3-5 (see page 145 correction), also K in Eq. 3-0.13 must be greater than 0

p.98 below Eq. 3.1-7, $1/\sqrt{2\pi}$ should be enclosed in parentheses for clarity

p.99 in Eq. 3-1.11, the 1st N in the parentheses should have a minus sign. Note that the Incomplete Gamma Function is defined for N an integer $>= 1$ and Y_b is a number greater than zero. In Eq. 3-1.12, the 1st bracketed series is finite, the last term is $(N-1)!/Y_b^{N-1}$. (The 1st bracketed series in Eq. 3-1.13 is correctly shown as infinite)

p.105 at the bottom, "Note 2" in the ANNOT column should read "SNDB"

pp.108, 118, 126 The upper left box of the card form should read "P_{fa}", as should step 42 on page 108

p.109 step 321 has the value $(2\beta^2-1)^2$ for case 4 or $(2\bar{x}+1)^2$ for case 5

p.110, 111 the example for this program only exercised Case 4 for N=1, which gives correct results. However, incorrect results will be obtained for N=2 to 9 since Register 1 returns the wrong value $(Yb/N\beta)$ instead of the desired value (Yb/β) from the Incomplete Gamma Function subroutine. Change steps 560, 561 to HIR 15. For N = 6, n' = 10^6, $\bar{x}$ = 10, old = .8674781097, new = .9924191005. Also, the exponent of β next to step 469 should be 3, not 2.

p.114 delete open parenthesis in Eq. 3-2.4 in front of ' a'. Although Eq. 3-2.5 is correct, it cannot be evaluated accurately for large n because the term $0.5^{1/n}$ approaches 1 and the difference from 1 approaches 0 due to the limited length of numbers in the calculator. By expanding this term in series, the expression in parenthesis becomes approximately $\ln(2)/n$, which can be evaluated accurately when n is larger (greater than 10^4), as shown in corrected program on page 135

p.116 step 3 b), delete overbar on x. Example 3-2.1 used a less accurate equation for "b" than is in the program so results are slightly different from results obtained with the program as listed in the book. Using the modified program on page 135, the value of Pd for NINT=1, n=1EE4, R/R0=.5 is 0.916368002, and for n = 10^{12}, R/R0=.4 is .9217425405

pp.119-120 as discussed above, a modified Program 3-2 is on page 135.

p.138 delete annotation: "t=0" at step 670

p.141 1st line, "c" should be upper case "C". Eq. 3-5.5 should read: $P_{FR}(M/N)=C(KP_{fa})/K$; 34rd line below Eq. 3-5.5, "alarm" should be "report"; 6th line, delete "with 64 filters"; 7th line, delete "/64". K in this case is the number of Doppler filters that may correlate with the 1st detection

p.142 for clarity, add parenthesis around $Y_{max}/2$ in Eq. 3-5.10 also, single-starred footnote is also footnote 1, the double-starred one is also footnote 2, and under eq. 3-5.8 should read "$Y^{max} = X_{99}/\bar{x}$

p.143 in Table 3-5.1, note that ρ must be grater than 1 and K must be greater than 0

p.145 in step 8 and 9, footnote should be "c", not "a"; in step 9 b),v) footnote should be "d," not "b"; step 10 b), footnote should be "e", not "c"; in step 8 the Log-normal parameter, ρ, must be greater than 1. To be compatible with the changes on p. 141, and revised Pgm 3-5, change step 11 as follows: step 11 b), enter KP_{fa}, step 11 c) divide by correlation window factor, K, VALUE is P_{FR}

p.149 the cards read at the bottom are Program 3-5, the value "28.2" in the COMMENTS column should be "28.8"

p.150 delete overbar on $\bar{x}_{99}$, REG 9

p.151 step 104, change x to $\bar{X}$

p.152 step 214 should be annotated "Y_{max}", not "sum". Change steps 264-267 for 0 to CLR RCL 02 RTN.

p.153 card form at top should have "3" in right-hand box. Change steps 433, 434 to 02 67, steps 467, 468 to 02 64, delete steps 488-493, change old step 497 to "x", insert STO 08 after old step 498, insert "÷ RCL 08 =" after old step 550, change old 550 to 03. Comments for SBR CE are now incorrect, P's are multiplied by K

p.166 in Eq. 4-2.8, x is the signal-to-noise for a 1 sq. meter target, and σ_{min} is also in sq. meters

p.171 the 3rd ANNOT entry should read "ERO", not "E_{RO}"; in Fig. 4-2.2, the curve labeled "σ^0=20dB should be $\sigma^0 = -20$dB; also, in Fig. 4-2.2, the point at 60 nmi should be plotted at 4dB not 8dB

p.173 in step 133, there should be a dot over E_R

p.178 in 2nd line of 2nd para. D_{BB} should be D_{BR}

p.189 the 5th line under Jammed Performance should have a slash mark ("/") between "power" and "bandwidth"; the units of R in Eq. 4-4.7 are nmi

p.204 note that poles must be within the unit circle for the filter to be stable. The program checks for this condition and stops if it occurs

p.208 change 1 to .0881 for Ex. 5-1.2 last entry and outputs RP, C1

pp.211-213 move 211 to 213, 212 to 211, 213 to 212

p.215 (old) add "*" at end of 1st line, add footnote: "*Response is normalized to response at fT=0 or 0.5, which may be found with Program 5-1."

p. 217 add: "note: Since G(fT) is normalized to unity at fT=0.5, the average gain and clutter attenuation calculated by this program do not agree with the usual definitions. To compare to these definitions, the gain at fT=0.5 must be multiplied into Eq. 5-2.6 and divided into Eq. 5-2.7. The Improvement Factor is therefore correct since the changes cancel.

p.224 The equations should be 5-3.1a and 5-3. 1b

p.225 the un-numbered equations at the top should be Eq. 5-3.2a and 5-3.2b

p.227 the author of Reference 4 is Shreve, not Shrove

p.234 in Eq. 5-4.1, the lower limits "i" and "k" should be interchanged

p.241 the small figure of FFT output should precede Ex. 5-4.3. The second plotted value should be at -2 dB, not -4 dB

p.245 FFT results for N complex = 16 or 32, or N real=32, 64, may be in error due to program error in re-ordering data. Correct by changing step 18 on card 3 from PRD to SUM.

p.251 next to last paragraph: note that the normalization does not account for the normalization of the weights. This could be accomplished by dividing the data by the square root of the sum of the squares of the weights, an exercise left to the reader.

p.255 in the 2nd line of Ex. 5-5.2, card 1 should be called the LOAD card

p.259 add after Eq. 5-6.3: "so that x_o is found from $x_o=\cosh(\cosh^{-1}(r)/(n-1))$"

p.260 insert "/N" before closing parenthesis in Eqs. 5-6.5, 5-6.6. The summations in these equations should run from K=1 to N, not 0 to N-1

pp.273-274 the notation of Eqs. 5-7.2, 5-7.5 is confusing; "t" should be "Bt", "n/B" should be "n", and "i/B" should be "i". Also note that Fn in Eq. 5-7.1 is twice the Fourier coefficient, Fi as defined in Eq. 5-7.3 and 5-7.4 due to the even symmetry

p.274 the sentence preceding Eq. 5-7.7 should read: "The...actual noise bandwidth, B_n, to the noise bandwidth, B, without weighting:"

p.279 steps 229, 230 should be 21, not 44

p.281 the second product of Eq. 5-8.8 cannot have n equal to m, so the notation n≠m below n=1 should be placed under the product sign

p.282 just above Eq. 5-8.10, "sum of the squares" should read "square of the sum", also, the quantity in braces on the right-hand side should be squared

p.283 Reference 6 is dated April 1962

pp.289-297 in the header, the program number should be 5-9, not 5-8

p.300 in Fig. 5-10.1, left-hand rectangle, P (0) = P/N

p.301 the paper of Reference 1 was presented at the 37th annual meeting, not the 40th.

p.308 two lines above Eq. 5-11.1, the equation referenced should be 5-2.8, not 5-2.7

p.310 in step 2, step 4 of Program 5-6 needs to be performed, as well as steps 1, 2, and 3

p.314 the Hamming weighting will be more accurate if steps 315-321 are changed to read "8575+/-+1". The weights and improvement factors in the examples will then be slightly different from those shown.

p.318 in Ex. 5-12.1, the Program 5-9 card should be banks 1 and 3, and the Program 5-12 card is bank 2 only. In the last paragraph, the reference is to Program 5-9, Ex. 5-9.1, the 1st gain value

p.327 under Eq. 6-1.4, L_R is defined as greater than unity, so the ratio should read "(power in/power out)"

p.349 the second author of Reference 17 is "Bernfeld", not "Gernfield"

pp.352-353 Ex. 6-3.2, the printouts for positive and negative K values are interchanged; however, Fig. 6-3.4 is correct. The output is corrected by the change to page 358

p.358 change step 136 from "-" to "+" to correct sign of K

p.362 in steps 4 and 5, the ranges are in units of τ_c

p.364 the contents of Register 8 should read "$frac(mi_i mj_j mk_k)$"

p.365 the annotation next to step 39 should read $m_3 \leq 1$. At the top, in the card format, C_1 should have an arrow and "M" to the right

p.371 change footnote to read: "*Always performs Int on display register, but does not perform indicated operation when hand-held"

p.381 in the table, row 6, column 5 should be an apostrophe, and column 6 is superscript x

p.391 add footnote: "*Lower case letters indicate a variable numeric value."

Correction to Program 3-1

6.		NINT	Case 4 gives correct results for the value of N given in the example. N = 1.
1.	06	N'	
6.	06	N	
25.89278591		YB	However, for N = 2 to 9 the program gives erroneous results because
4.		CASE	Register 1 returns the wrong value from the Incomplete Gamma Function subroutine.
0.5		X	Using H5 corrects this problem.
.0001445407		PD	
1.		x̄	
0.001998722		PD	
3.		x̄	original results
.1245432812		PD	
10.		x̄	
.8674781097		PD	
10.		x̄	
.9924191005		PD	corrected result, agrees with Fehlner, Fig. 43

Correction to Program 3-2

240	76	LBL	300	95	=	360	43	RCL	420	00	00
241	14	D	301	23	LNX	361	02	02	421	62	62
242	32	X:T	302	65	×	362	95	=	422	42	STD
243	25	CLR	303	02	2	363	44	SUM	423	04	04
244	43	RCL	304	94	+/-	364	05	05	424	91	R S
245	16	16	305	95	=	365	43	RCL	425	32	X:T
246	10	E'	306	34	ГX	366	03	03	426	01	1
247	55	÷	307	42	STD	367	49	PRD	427	04	4
248	01	1	308	11	11	368	02	02	428	71	SBR
249	00	0	309	35	1/X	369	43	RCL	429	00	00
250	95	=	310	42	STD	370	01	01	430	62	62
251	22	INV	311	00	00	371	32	X:T	431	32	X:T
252	28	LOG	312	65	×	372	65	×	432	01	1
253	71	SBR	313	43	RCL	373	48	EXC	433	32	X:T
254	02	02	314	04	04	374	08	08	434	71	SBR
255	71	71	315	22	INV	375	32	X:T	435	03	03
256	81	RST	316	49	PRD	376	85	+	436	07	07
257	00	0	317	00	00	377	32	X:T	437	95	=
258	00	0	318	95	=	378	85	+	438	32	X:T
259	76	LBL	319	22	INV	379	01	1	439	03	3
260	15	E	320	77	GE	380	95	=	440	04	4
261	32	X:T	321	03	03	381	32	X:T	441	71	SBR
262	25	CLR	322	26	26	382	03	3	442	00	00
263	04	4	323	29	CP	383	22	INV	443	62	62
264	04	4	324	35	1/X	384	28	LOG	444	98	ADV
265	71	SBR	325	22	INV	385	77	GE	445	92	RTN
266	00	00	326	86	STF	386	03	03	446	76	LBL
267	62	62	327	01	01	387	46	46	447	19	D'
268	61	GTO	328	42	STD	388	43	RCL	448	32	X:T
269	02	02	329	03	03	389	04	04	449	25	CLR
270	53	53	330	42	STD	390	75	-	450	03	3
271	65	×	331	02	02	391	43	RCL	451	05	5
272	02	2	332	02	2	392	11	11	452	06	6
273	95	=	333	49	PRD	393	95	=	453	03	3
274	34	ГX	334	00	00	394	33	X²	454	03	3
275	42	STD	335	35	1/X	395	94	+/-	455	05	5
276	04	04	336	32	X:T	396	22	INV	456	00	0
277	82	HIR	337	42	STD	397	23	LNX	457	01	1
278	17	17	338	05	05	398	34	ГX	458	71	SBR
279	32	X:T	339	00	0	399	65	×	459	00	00
280	04	4	340	42	STD	400	43	RCL	460	62	62
281	22	INV	341	01	01	401	05	05	461	33	X²
282	28	LOG	342	42	STD	402	55	÷	462	33	X²
283	77	GE	343	06	06	403	02	2	463	35	1/X
284	02	02	344	42	STD	404	55	÷	464	61	GTO
285	92	92	345	08	08	405	32	X:T	465	02	02
286	02	2	346	43	RCL	406	87	IFF	466	53	53
287	23	LNX	347	00	00	407	01	01			
288	55	÷	348	44	SUM	408	04	04			
289	61	GTO	349	01	01	409	13	13			
290	02	02	350	43	RCL	410	94	+/-			
291	98	98	351	01	01	411	85	+			
292	01	1	352	65	×	412	01	1			
293	75	-	353	43	RCL	413	92	RTN			
294	93	.	354	06	06	414	76	LBL			
295	05	5	355	48	EXC	415	13	C	REG 07		
296	22	INV	356	05	05	416	32	X:T	NOT USED		
297	45	YX	357	42	STD	417	01	1			
298	01	1	358	06	06	418	03	3			
299	32	X:T	359	85	+	419	71	SBR			

Appendix D

PROGRAM PATPLOT/1

This program is a revised version of program PATPLOT to be used with the Radio Shack CGP-115 color graphic printer. This printer is actually a 4-pen plotter, with a resolution of 0.2mm in both dimensions. The plotter is relatively low cost (under $200), and can be driven with either a Centromics (parallel) interface or an RS-232C (serial) interface. The HP-86A Centronics interface, normally used with a line printer, is compatible with this plotter. The plotting commands use several escape codes and single-letter commands, making programming quite easy. Both text and graphic modes are available. This program is enhanced over the original PATPLOT in that the plot subroutine is written as a general purpose program for use in other programs. Multiple overlaid plots are also possible. Resolution is, of course, much better than the line printer plot.

```
10  ! Pgm PATPLOT/1, rev 2/11/84, by W. Skillman
20  ! Derived from Program 1-2, Radar Calc..
30  !
40  ! Improvements vs. Radar Calc...
50  !    Up to 10X Angle Resolution (2.5 deg vs 0.25)
60  !    0 to -60 dB Gain Scale (vs 0 to -40)
70  !    Line plot (Radio Shack Color Graphic Printer CGP-115, a 4-pen plotter)
80  !
90  ! Important variables:
100 !    CTH..COS(TH) array
110 !    D....element spacing
120 !    F....Gain (dB) array
130 !    LAM..wavelength
140 !    M....# of elements
150 !    STH..SIN(TH) array
160 !    TH...angle from array normal(deg)
170 !    THS..scan angle from   "
180 !
190 DIM Y$[60],X$[73],F(721),STH(721),CTH(721),T$[25],S$[25]
210 !
220 T$="ARRAY ANTENNA PATTERN" @ S$="Unweighted Aperture"
230 Y$="Relative Gain (dB)" @ X$="Angle (deg)"
275 W$="Please wait while "
280 CLEAR @ DEG @ PRINTER IS 701
285 DISP "Enter today's date (any comma-less format)";@ INPUT DA$
290 DISP "Enter point spacing (deg) (>=.25,default=1)";@ INPUT R$
300 DDX,I=1 @ IF R$#"" THEN DDX=VAL (R$)
305 IF DDX<.25 THEN DISP "Use spacing>=.25" @ BEEP @ GOTO 290
310 II=180/DDX+2 @ CLEAR @ DISP W$;"I store sines and cosines"
320 FOR TH=-90 TO 0 STEP DDX @ CTH(II-I),CTH(I)=COS (TH)
330 STH(I)=SIN (TH) @ STH(II-I)=-STH(I) @ I=I+1 @ NEXT TH @ BEEP 20,50 @ CLEAR
340 !
```

```
350 ! **********   MAIN PROGRAM   **************************
352 MFLAG=0 @ DISP "Select Grid type" @ DISP
353 DISP "0  Tics (+)"
354 DISP "1  Dashed lines"
355 DISP "2  No Grid" @ DISP
356 DISP "Enter #";@ INPUT R@ GFLAG=R
357 IF R<0 OR R>2 THEN DISP "Use # 0-2" @ BEEP @ GOTO 356
360 YT=0 @ YB=-60 @ DX,DY=10 @ XL=-90 @ XR=90 ! Grid parameters
365 GOSUB 430 ! enter antenna parameters
370 GOSUB 9500 ! plot pattern
380 DISP "Select next operation" @ DISP
382 DISP "1  Overlay plot" @ DISP "2  New Plot"
383 DISP "3  Disc Menu" @ DISP
384 DISP "Enter #";@ INPUT R@ IF R=3 THEN CHAIN "Autost"
386 IF R<1 OR R>3 THEN DISP "Use # 1-3" @ BEEP @ GOTO 384
390 ON R GOTO 392,396
391 !
392 MFLAG=MFLAG+1 @ GOTO 365
393 !
396 MFLAG=-1 @ GOSUB 10260 @ GOTO 352
420 !
430 ! **********   Enter antenna parameters   ***************
440 DISP "Enter element spacing and wavelength (same units)";
445 INPUT D,LAM@ IF D<= 0 OR LAM<= 0 THEN DISP "Use #s>0" @ BEEP @ GOTO 440
450 IF D<LAM/2 THEN DISP "Use spacing >= wavelength/2" @ BEEP @ GOTO 440
460 DISP "Enter # of antenna elements";@ INPUT M
470 IF M<= 0 THEN DISP "Use # >0" @ BEEP @ GOTO 460
480 DISP "Enter desired scan angle (deg)";@ INPUT THS@ SAS=SIN (THS)
490 IF ABS (THS)>90 THEN DISP "Use angle less than +/-90" @ BEEP @ GOTO 480
500 DISP W$;"I calculate gains for the plot" @ Y2=-800 @ XT=-75
505 IF MFLAG THEN 525
510 CR$=CHR$ (13) @ PRINT CHR$ (17);CR$;CHR$ (18);CR$;"S1"
520 XT=75 @ Y2=0 @ PRINT FNM$(100,0);"P";T$;CR$;FNM$(250,-20);"P";DA$
525 PRINT FNM$(XT,Y2-40);"S1";CR$;"Q0"
530 PRINT "PSpacing=";D;",  Wavelength=";LAM
535 PRINT FNM$(XT,Y2-60);"PNo. of elements=";M;", Scan angle=";THS
540 FOR I=1 TO II-1 @ GOSUB 560 @ NEXT I @ RETURN
550 !
560 ! *************   CALC GAIN FOR UNIFORM ILLUM   **************
570 u=180*D/LAM*(STH(I)-SAS) @ R=1
580 Su=SIN (u) @ IF Su THEN R=SIN (M*u)/M/Su
590 E=R*(1+CTH(I))/2 @ E=MAX (EPS ,ABS (E)) @ F(I)=20*LGT (E) @ RETURN
600 !
9500 ! *************   Plotter subroutine   ********************
9510 !
9520 ! Plot sub revised 2/11/84, W. Skillman, for Radio Shack CGP-115
9530 !   4 1/2 inch plotter.  Plot rotated 90 deg, X (horizontal) axis
9540 !   along length of paper.  There are 480 graphic units across page.
9550 !   Graph area is approx 360 x 720 gu's, scaled so integer # of
9560 !   gu's per segment (between tics or grid lines)
```

```
9570 !    Points are equispaced in X. Multiple plots require point array,F,
9580 !    to be reloaded between sub calls. MFLAG used to annotate multiple
9585 !    plots. Y-axis labels, X,Y-axis legends, titles centered
9590 !
9600 ! Input variables for plot sub:
9610 !    DDX.....Point X-spacing in input units
9620 !    DR$...Date, or other notation for lower right corner
9630 !    DX,DY...Grid or tic and label spacing in X,Y input units
9640 !    F(i)....Y value of input point, i=1 at XL, last point at XR
9650 !    GFLAG...=0, +tics, =1, dashed line grid =2, frame tics only
9660 !    LT.....Line type, 0-15, 0 is solid, used for 1st plot, incremented
9670 !            by 2 each succeeding plot
9680 !    MFLAG..Multiple plot counter, =# of plot-1
9682 !    S$....Subtitle <63 char
9685 !    T$...Title,<42 char
9690 !    X$,Y$...X and Y axis legends
9700 !    XL,XR...Left and right X limits, input units, XR>XL
9710 !    YT,YB...Top and bottom Y limits, input units,YT>YB
9720 !
9730 ! Internal variables:
9740 !    DXP...DDX, gu's
9750 !    NTi...# of tic segments on i axis
9760 !    NUi...# of graph units/segment on i axis
9770 !    XRU...XR in gu's
9780 !    YTU...YT in gu's
9790 !
9800 IF MFLAG THEN 10220 ELSE PRINT CHR$ (17) ! Text mode unless multiple plots
9810 IF YT<= YB THEN DISP "Use YT>YB" @ BEEP @ PAUSE
9820 IF XR<= XL THEN DISP "Use XR>XL" @ BEEP @ PAUSE
9830 IF DX<= 0 OR DY<= 0 OR DDX<= 0 THEN DISP "Use DX,DY,DDX >0" @ BEEP @ PAUSE
9840 CR$=CHR$ (13) @ PRINT CHR$ (18);CR$;"I" ! graphics mode, origin at L edge
9850 !
9860 DEF FNU$(X,Y) = VAL$ (X)&","&VAL$ (Y)&CR$ ! Utility
9870 DEF FNM$(X,Y) = "M"&FNU$(X,Y) !          Move-absolute
9880 DEF FNR$(X,Y) = "R"&FNU$(X,Y) !          Move-relative
9890 DEF FND$(X,Y) = "D"&FNU$(X,Y) !          Draw-absolute
9900 DEF FNS$(X) = "S"&VAL$ (X)&CR$ !         Char size
9910 DEF FNJ$(X,Y) = "J"&FNU$(X,Y) !          Draw-relative
9920 DEF FNJR$(A,B,C,D) = FNJ$(A,B)&FNR$(C,D) ! Draw/Move-relative
9930 DEF FNX$(A,B,C) = "X"&VAL$ (A)&","&FNU$(B,C) ! Axis
9940 !
9950 NTY=(YT-YB)/DY @ NTX=(XR-XL)/DX @ NUY=INT (360/NTY+.5)
9960 NUX=INT (720/NTX+.5) @ YTU=NTY*NUY @ XRU=-(NTX*NUX)
9962 LEY=MAX (LEN (VAL$ (YB)),LEN (VAL$ (YT)))
9964 IF YT>= 0 OR YB>= 0 THEN LEY=LEY+1
9966 PL=-148+12*LEY ! Sets left edge of Y-legend, labels
9970 PRINT "Q0";CR$;"L0";CR$;FNM$(230-6*LEN (Y$),PL+10);FNS$(1);"P";Y$;CR$;"Q1"
9980 YY=43 @ LT=0 @ FOR Y=YB TO YT STEP DY ! label Y-axis
9990    PRINT FNM$(YY,PL);"P";Y @ YY=YY+NUY
10000 NEXT Y
```

```
10002 PRINT FNM$(456,-510+9*LEN (T$));FNS$(2);"P";T$ ! Title
10003 PRINT FNS$(1);FNM$(430,-510+6*LEN (T$));"P";S$ ! Subtitle
10005 PRINT FNM$(0,-510+6*LEN (X$));"P";X$ ! X-axis legend
10010 PRINT FNS$(0);FNM$(0,-870+6*LEN (DA$));"P";DA$ ! date
10020 XX=-143+XRU @ FOR X=XR TO XL STEP -DX ! label X-axis
10030    PRINT FNM$(25,XX);"P";X @ XX=XX+NUX
10040 NEXT X
10050 PRINT FNM$(50,-150);"I" @ IF GFLAG=1 THEN 10080 ! reset origin
10060 PRINT FNX$(1,NUY,NTY);FNX$(0,-NUX,NTX);FNX$(1,-NUY,NTY);FNX$(0,NUX,NTX)
10065 PRINT "Q0" @ Q=-4-NUX @ IF GFLAG THEN 10220 ELSE 10150 ! Frame with axes
10070 !
10080 Y1,X1=0 @ X2=YTU @ Y2=XRU @ GOSUB 10480 ! Frame graph area with line
10090 K=XRU @ PRINT FNM$(NUY,0);"L4" ! Vertical dashed lines
10100 FOR I=1 TO NTY-1 @ PRINT FNJR$(0,K,NUY,0) @ K=-K @ NEXT I
10130 K=YTU @ PRINT FNM$(X,-NUX)
10140 !
10150 FOR I=1 TO NTX-1 ! Horiz dashed lines or +
10160    IF GFLAG THEN PRINT FNJR$(K,0,0,-NUX) @ K=-K @ GOTO 10200 ELSE P=-3+NUY
10170    FOR J=1 TO NTY-1 @ PRINT FNM$(P,Q);"P+" @ P=P+NUY @ NEXT J @ Q=Q-NUX
10200 NEXT I @ LT=0
10210 !
10220 ! PRINT "L";LT @ GOSUB 10510 ! Label line type (not used this pgm)
10225 GOSUB 10440 @ I=2 @ DXP=NUX*DDX/DX ! Draw line graph
10230 FOR X=-DXP TO XRU STEP -DXP @ GOSUB 10290 @ I=I+1 @ NEXT X @ LT=LT+2
10260 IF MFLAG<0 THEN PRINT FNM$(0,-800);FNS$(1);"A"
10270 RETURN
10280 !
10290 ! ************** Draw/Move to next point  **************
10300 !
10310 YL=Y @ Y=F(I) @ XX=X @ C$="M" @ IF YY=YT THEN 10410
10330 IF YY=YB THEN 10400
10340 C$="D" @ IF Y>YT THEN YY=YT @ GOSUB 10380 @ GOTO 10460
10350 IF Y<YB THEN YY=YB @ GOSUB 10380 @ GOTO 10460 ELSE YY=Y
10360 PRINT C$;YTU-(YT-YY)*NUY/DY;",";XX @ RETURN
10370 !
10380 XX=DXP*(1-ABS ((YY-YL)/(YL-Y)))+X @ GOSUB 10360 @ XX=X @ RETURN
10390 !
10400 IF Y<YB THEN 10360 ELSE 10420
10410 IF Y>YT THEN 10360
10420 GOSUB 10380 @ C$="D" @ YY=Y @ GOTO 10360
10430 !
10440 YY,Y=F(1) @ XX=0 @ IF Y>YT THEN YY=YT
10450 IF Y<YB THEN YY=YB
10460 C$="M" @ GOTO 10360
10470 !
10480 ! ********** Sub to draw box  ****************************
10490 PRINT FNM$(X1,Y1);FND$(X2,Y1);FND$(X2,Y2);FND$(X1,Y2);FND$(X1,Y1)
10495 RETURN
```

www.ingramcontent.com/pod-product-compliance
Lightning Source LLC
Chambersburg PA
CBHW081539220326

41598CB00036B/6493